Everyday Mathem

The University of Chicago Scho

5 - Minute Math®

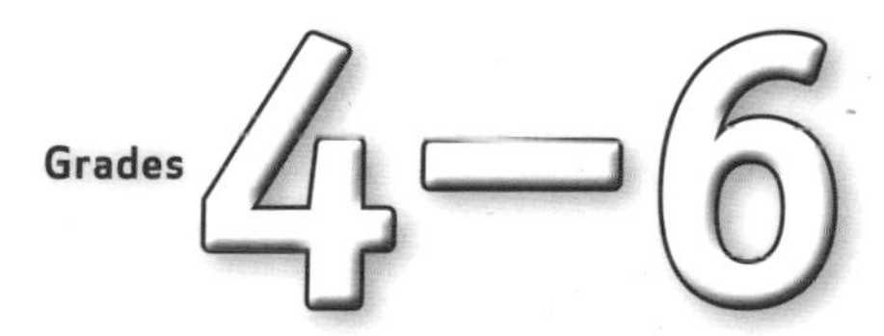

Chicago, IL • Columbus, OH • New York, NY

everyday**math**.com

Send all inquiries to:
McGraw-Hill Education
P.O. Box 812960
Chicago, IL 60681

ISBN: 978-0-07-657723-1
MHID: 0-07-657723-6

Printed in Mexico.

1 2 3 4 5 6 7 8 9 DRN 17 16 15 14 13 12 11

Developed by
Sheila Sconiers
Mary Kay Dyer
Sue Rasala
Kathleen Snook

Technical Art
Diana Barrie

Photo Credits
Cover (l)John W Banagan/Iconica/Getty Images, (c)Pier/Stone/Getty Images, (r)Digital Vision/Getty Images.

CONTENTS

Introduction v
Easy Activities 1
Numeration 1
Operations 19
Data 34
Probability 42
Measurement 48
Geometry 56
Algebra 66
Moderate Activities 79
Numeration 79
Operations 95
Data 114

Probability 125
Measurement 131
Geometry........................ 139
Algebra 150
Difficult Activities 165
Numeration....................... 165
Operations........................ 182
Data............................... 198
Probability 206
Measurement 212
Geometry........................ 220
Algebra 231
List of Activities by Page 247

Using *5-Minute Math*

PURPOSE: The *5-Minute Math* activities are intended to provide additional practice and review during occasional 5- to 10-minute "down" times throughout the school day. Many of the activities can be used as prompts for writing practice. This book is a rich source of Open Response and Extended Response opportunities.

FORMAT: The color-coding indicates the degree of difficulty of the activities in each section (Easy, light blue; Moderate, gray; Difficult, dark blue). Within each section, the activities are grouped by strand.

Each activity begins by introducing a topic and posing a question. This is followed by 2 or 3 progressively more challenging questions so that each activity can be tailored to fit different needs and time slots.

For your convenience, answers are provided for most problems. The answers are in parentheses following the question and are printed in blue to distinguish them from the rest of the text. If no answer appears, many answers are possible. Note that answers also appear in the tables. These should not be included when the table is written on the board and students are asked to complete it.

CONTENT: In many cases the content mirrors the content of *Fourth, Fifth,* and *Sixth Grade Everyday Mathematics*®. However, we have also included problems and vocabulary that may occur on a variety of standardized tests at these grade levels.

PREPARATION: For many activities, students will need slates or scratch paper, and occasionally, a few calculators will be useful. Otherwise no special preparation is needed.

Fractions

Set up: Write $\frac{1}{2}$ on the board.

1. **Tell me a fraction that is equivalent to $\frac{1}{2}$.** (Example: $\frac{2}{4}$)
   Write the fraction on the board.
   Another one? Repeat until you have written about 20 fractions equivalent to $\frac{1}{2}$.

2. **Come to the board and loop together two fractions that have a sum of one.**
   Continue to invite students to loop together pairs of fractions until the class concludes that any two fractions on this list will add up to one.

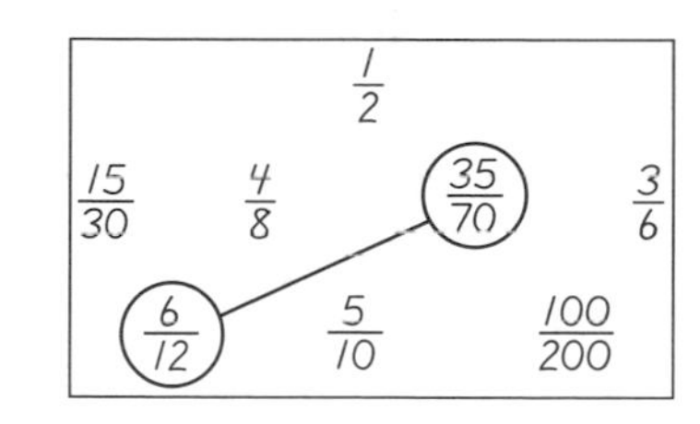

3. **Now let's use a different color chalk and loop fractions that add up to $2\frac{1}{2}$.** (any 5 fractions)

Reusable: Repeat the activity using fractions equivalent to $\frac{1}{3}$.

Numeration

Powers of Ten

Set up: Draw a three-column chart labeled as shown below and write 10^2 in the first row, first column.

Power of 10	Standard Notation	Multiply by 10s
10^2	(100)	(10∗10)

1. **This number is a power of 10. How is this number read?** ("ten squared")
   Tell me how to write "ten squared" or "ten to the second power" in standard notation. (100)
   How would we multiply by 10s to get this number? (10∗10)
   Write responses in the second and third columns.
   What is the 2 in "ten to the second" called? What is the 10 called?
   (exponent, base)

continued

2. Write 1,000 in the second row, second column of your chart.
   Don't call out, but show me by holding up fingers: What exponent is used to write 1,000 as a power of 10? (3)
   Survey student answers and write 10^3 in the first column of your chart.
   How is this read? ("ten cubed")
   How should we write 1,000 as a product of 10s? ($10 * 10 * 10$)
   Complete the second row of the chart.

3. **Now hold up fingers to show what exponent is used to write 10,000.** (4)
   Tell me how to write 10,000 as a product of 10s. ($10 * 10 * 10 * 10$)
   Add and complete the next row of the chart.

4. **What patterns do you notice in the chart?**

Reusable: Draw a 3-column chart with the same column labels and one of the above examples filled in. Repeat this activity by entering the chart at different points: For example, enter $10*10*10*10*10$ in the third column, enter 1,000,000 in the second column, or enter 10^7 in the first column. Have students complete each row.

Numeration

Roman Numerals: I–XXX

Set up: Draw the chart below on the board.

1. **Roman Numerals are an old system of writing numbers. You sometimes see them on clocks. What do you think V equals? How do I write 3?** (5, III) **Tell me how to fill in the rest of the blanks.**

2. Extend the chart.
   Look at 9. How do you think 10 is written in Roman Numerals? Eleven? (X, XI)
   What pattern do you notice about the placement of I with another character? (I's on the left decrease the value. I's on the right increase the value.)

   Tell me how to fill in 12 and 13. (XII, XIII)
   Fourteen is written as XIV. Tell me how to fill in 15 through 18.
   (XV, . . ., XVIII)

Roman Numerals	Standard Notation
I	1
(II)	2
(III)	3
IV	4
(V)	5
VI	6
(VII)	7
(VIII)	8
IX	9

continued

3. **Twenty is written as XX. Tell me how to write 19 and 21.** (XIX, XXI)
   Let's complete the column of Roman Numerals to 30.
   Describe the patterns that you see.

Roman Numerals: I–XXX (Review)

Set up: The board is needed. Students need slates or paper.

1. **Quickly write down the Roman Numerals 1 through 10.**
   How many different symbols are used? What are they? (3; I, V, X)
   In some numbers symbols are repeated. Which number uses the most symbols? Repeated symbols count separately. (8 or VIII)
   Which numbers use just two symbols? (2, 4, 6, 9 or II, IV, VI, IX)

2. **How can you quickly change your list to the Roman Numerals for 11 through 20? 21 through 30?** (Add an X to the left side, add two Xs to the left side) **Which Roman Numeral for the numbers from 1 to 30 uses the most symbols? Repeated symbols count separately.** (28 or XXVIII)
   What is the largest Roman Numeral from 1 to 30 that can be written with just two symbols? (20 or XX)

3. Write IX and XI on the board.
   What are the values of these Roman Numerals? (9, 11)
   Find other pairs of Roman Numerals from 1 to 30 that use exactly the same symbols. Write them on your paper. (Example: 4 and 6, or IV and VI)

Using Exponents

Set up: Write 100 and 10 * 10 on the board.

1. **Why is 100 or 10 * 10 called a square number?** (Example: A square array of 10 rows of 10 dots each equals 100 dots.)
   What is another way of writing "10 times 10"? (10^2)
   Write it on the board.
   Read each notation. ("ten times ten," "ten squared")

2. **How is "five squared" written?** (5^2)
   Write it on the board.
   What is the 2 called in "five squared"? (exponent) **What does it indicate?** (Example: That five is multiplied by itself) **What does "five squared" equal?** (25)

3. **Name other square numbers less than 100 and tell me how to write each number using an exponent.** (Example: 64, 8^2)

4. Write $1{,}000 = 10^3$ on the board.
   Why is this true? How is this read? ("one thousand equals ten cubed")
   How do you write "two to the fourth"? What does this equal? (2^4, 16)

Numeration

Numeration

Fractions: $\frac{1}{2}$

Set up: The board is needed. Students need slates or paper.

1. Quickly survey the class and choose two students that obviously share an attribute and three who do not. You might look for shirt color, shirts with buttons or no buttons, or style of shoe.
   Two out of these 5 students have shirts with buttons. Write the fraction of students who have shirts with buttons. ($\frac{2}{5}$)
   Write down the fraction of students that do not have shirts with buttons. ($\frac{3}{5}$)
   What is the denominator? What is the numerator? (5, 3)
   Which fraction is more than half? ($\frac{3}{5}$)
   Ask two more students to join the group.
   Now what fraction of the group has shirts with buttons?
   Is this more or less than half the group?

2. On the board, write down the names of a group of students where one-half of the group is wearing sneakers. Use a small group, say four or six students.
   How many students are wearing sneakers out of the total group?
   (Example: $\frac{3}{6}$ or $\frac{1}{2}$)
   Add some students to the group but be sure that one-half of the group is still wearing sneakers.
   Describe our new group by saying the number of students wearing sneakers out of the total.
   Write the corresponding fraction.

3. **If there are 20 students at a bus stop and half of them have gym bags, how many have gym bags?** (10)
   If a little less than half of the 20 students get on the first bus, how many are still waiting? (about 11 students)
   If nine students got on the second bus and just over half of them are girls, how many boys got on the bus? (4 boys)

Numeration

Divisibility Rules for 2

Set up: None

1. **If we divide a number by 2 and the remainder is zero, then that number is said to be divisible by 2.**
   Is the number of students in the class divisible by 2? What about the number of days in a week? (no) **The number of days in this month?**
   Is the number of months in a year divisible by 2? (yes)

2. **What do we mean when we say a number is even? Or odd?**

3. **I'm going to read some statements. If a statement is true show me a thumbs-up. If it's false show me a thumbs-down.**
   32 is an odd number. (false)
   41 is an even number. (false)
   Finish this sentence: Numbers that have 0, 2, 4, 6, or 8 in the ones place are _____. Be sure it's true. (even)

4. **Explain how you know whether the number of days in this year is divisible by 2.**

Fractions: <1, $=1$, >1

Set up: Write 2 and 4 on the board. Students need slates or paper.

1. **Using only these numbers, write down as many fractions as you can.** After a minute, write the fractions on the board. ($\frac{2}{4}$, $\frac{4}{2}$, $\frac{2}{2}$, $\frac{4}{4}$)
   Which fraction is the largest? ($\frac{4}{2}$) **Which fraction is the smallest?** ($\frac{2}{4}$)
   Which fractions are equivalent? ($\frac{2}{2}$ and $\frac{4}{4}$)

2. **Now expand your list of fractions by writing all the fractions you can using 2, 3, and 4. Sort the fractions into three groups: less than 1, equal to 1, and greater than 1.** ($\frac{2}{3}$, $\frac{2}{4}$, and $\frac{3}{4}$, $\frac{2}{2}$, $\frac{3}{3}$, and $\frac{4}{4}$, $\frac{3}{2}$, $\frac{4}{2}$, and $\frac{4}{3}$)
   Let's write the fractions in order. Which is the smallest? Next? ($\frac{2}{4}$, $\frac{2}{3}$, . . .)
   Which fractions are improper fractions? ($\frac{2}{2}$, $\frac{3}{3}$, $\frac{4}{4}$, $\frac{3}{2}$, $\frac{4}{2}$, $\frac{4}{3}$)
   Which fractions can be renamed as a mixed number? ($\frac{3}{2}$, $\frac{4}{2}$, $\frac{4}{3}$)

3. **What do you know about the numerator and denominator of a fraction that is less than 1?** (The denominator is greater than the numerator.)
   A fraction equal to 1? (The denominator and numerator are equal.)
   A fraction greater than 1? (The denominator is less than the numerator.)

Reusable: Repeat this activity with other sets of numbers, such as 3, 5, and 10.

Whole Number Place Value

Set up: Write the numbers 12; 123; 1,234; 12,345; and 123,456,789 on the board in a column. Students need slates or paper.

1. **Use expanded notation to show the place value of each digit in the numbers shown on the board.** (Example: 123 = 100 + 20 + 3)

2. **Write the name for each of the numbers in words.** (Example: 1,234 = one thousand two hundred thirty-four)

Whole Number Factors

Set up: Write 12, 24, and 36 on the board. As an example, draw a multiplication/division Fact Triangle with 4, 6, and 24. Students need slates or paper and could use graph paper for Problem 2.

1. **Draw a picture of all the multiplication/division Fact Triangles that would have the products shown on the board.**

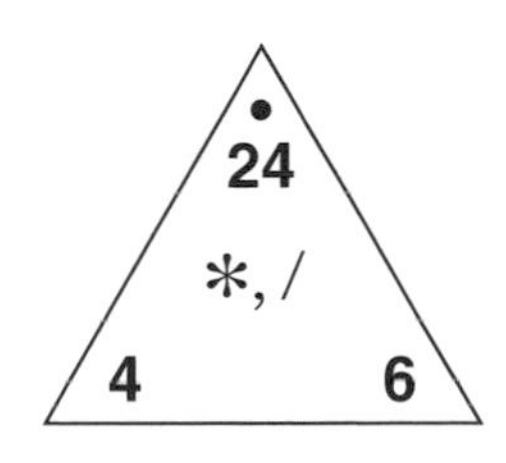

2. **Draw an array to represent each of the Fact Triangles that you drew.** (See art at right for an example.)

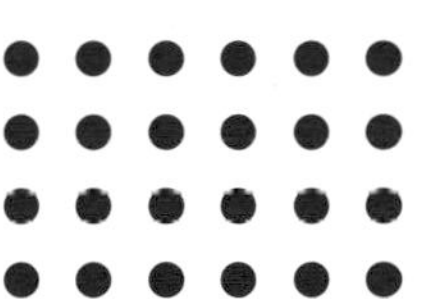

Decimal Place Value

Set up: Write the numbers 2.54, 1.3, 0.56, 1.23, 0.09, 2.67, and 0.1 on the board. Also draw the following example:

3.12 = □□□|▪▪

Students need slates or paper.

1. **Use large squares, line segments, and small squares to show each of the numbers given on the board.**
   The large square stands for the base-10 *flat* and represents 1 whole; the line stands for a *long*, which is $\frac{1}{10}$ of the flat; and the small square stands for the *small cube*, which is $\frac{1}{100}$ of the flat.

2. **Now put these numbers in order from smallest to largest.** (0.09, 0.1, 0.56, 1.23, 1.3, 2.54, 2.67)

Whole Number Rounding

Set up: Draw the first chart on the board. While students are working on the first chart, you can draw the second chart on the board. Students need slates or paper.

1. Draw this chart on your paper. Round the numbers in the chart to the places given.

	Nearest 1,000	Nearest 100	Nearest 10
36,555	(37,000)	(36,600)	(36,560)
609,909	(610,000)	(609,900)	(609,910)
999,999	(1,000,000)	(1,000,000)	(1,000,000)

2. Now round these numbers to the places indicated.

	Nearest tenth	Nearest hundredth
0.467	(0.5)	(0.47)
8.049	(8.0)	(8.05)
3.095	(3.1)	(3.10)

Numeration

Fractional Parts

Set up: Draw the chart below on the board. Students need slates or paper.

1. **Draw this chart on your paper. We want to compare the value of each coin to $1, $5, and $10. For example, what part of $1 is a quarter? Write your answers in fraction and decimal form. One row is done for you.**

	$1	$5	$10
quarter	$\frac{25}{100}$ 0.25	$\frac{25}{500}$ 0.05	$\frac{25}{1,000}$ 0.025
dime	($\frac{10}{100}$ 0.1)	($\frac{10}{500}$ 0.02)	($\frac{10}{1,000}$ 0.01)
nickel	($\frac{5}{100}$ 0.05)	($\frac{5}{500}$ 0.01)	($\frac{5}{1,000}$ 0.005)
penny	($\frac{1}{100}$ 0.01)	($\frac{1}{500}$ 0.002)	($\frac{1}{1,000}$ 0.001)

2. **Now let's compare whole dollar amounts.**
   Draw a new chart to compare $1, $2, $3, $4, and $5 (rows) to $5, $10, $20, and $100 (columns).

Equivalent Fractions

Set up: Write the numbers shown below on the board. Students need slates or paper.

1. **Write three equivalent fractions for each of the numbers shown on the board.** (Example: $\frac{1}{5} = \frac{2}{10} = \frac{3}{15} = \frac{4}{20}$)

$\frac{1}{5}$	$\frac{1}{4}$	$\frac{1}{3}$	$\frac{1}{2}$
$\frac{2}{5}$	$\frac{2}{4}$	$\frac{2}{3}$	$\frac{2}{2}$
$\frac{3}{5}$	$\frac{3}{4}$	$\frac{3}{3}$	$\frac{3}{2}$
$\frac{4}{5}$	$\frac{4}{4}$	$\frac{4}{3}$	$\frac{4}{2}$
$\frac{5}{5}$	$\frac{5}{4}$	$\frac{5}{3}$	$\frac{5}{2}$

2. **What patterns do you notice?**

Whole Number Place Value

Set up: Draw the first chart on the board. While students are working on the first chart, you can draw the second chart on the board. Students need slates or paper.

1. Copy and complete this chart. Let's do the first one together.

	10 less	10 times smaller
120	110	12
2,340	(2,330)	(234)
164,000	(163,990)	(16,400)

2. Copy and complete this chart.

	100 less	100 times smaller
1,500	(1,400)	(15)
93,400	(93,300)	(934)
2,280,000	(2,279,900)	(22,800)

Multiplication Estimation

Set up: Students need slates or paper. You may want students to have calculators.

1. **How many student arms and legs are in this classroom? How did you determine your answer?** (number of students * 4)

2. **A fourth grader usually has 28 teeth. Estimate the total number of students' teeth in this classroom.** (number of students * 28) **Why is this an estimate?** (Example: We didn't actually count the teeth and some students may not have exactly 28.)

3. **A baby has about 300 separate bones. Estimate the total number of bones the students in this class had when you were all infants.** (number of students * 300)

4. **As a person grows, his or her bones fuse together. Does anyone know what "fuse together" means? Because of this, an adult has about two-thirds the number of bones of a child. Estimate the total number of bones the students in this class will have when you are adults.** (number of students * 300 * $\frac{2}{3}$)

Source: Beth Rowan and Curtis Slepian (Eds.), *Time for Kids Almanac 2005* (Needham, Mass.: Pearson Education, 2004), 107.

Division Word Problems

Set up: The board is needed. Students need slates or paper.

1. **Using a number greater than 20, make up a division problem with a divisor of 6 and a remainder of 4.** (Example: 22 / 6 ⟶ 3 R4)
   List some responses on the board.

2. **Make up two different word problems that go with your division problem.**
   Have students share problems.

3. **List all the numbers less than 50 that can be used to make up a division problem with a divisor of 6 and a remainder of 4.** (10, 16, 22, 28, 34, 40, 46)
   Do you notice a pattern? (Example: Each number is 6 more than the last.)

Reusable: Repeat this activity, asking students to make up division problems with a divisor of 3 and remainder of 2 or a divisor of 5 and remainder of 2.

Whole Number Addition

Set up: Write the digits 0, 1, 2, 3, and 4 on the board. Students need slates or paper.

1. **Use each of the digits 0, 1, 2, 3, and 4 in an addition problem to make the largest sum possible.** (4,321 + 0 = 4,321 or 4,320 + 1 = 4,321)

2. **Use each of the digits 0, 1, 2, 3, and 4 in an addition problem to make the smallest sum possible.** (0 + 1 + 2 + 3 + 4 = 10)

3. Write the numbers 5, 6, 7, 8, and 9 on the board.
   Use each of the digits 5, 6, 7, 8, and 9 in an addition problem to make the largest sum possible. (9,876 + 5 = 9,881 or 9,875 + 6 = 9,881)

4. **Use each of the digits 5, 6, 7, 8, and 9 in an addition problem to make the smallest sum possible.** (5 + 6 + 7 + 8 + 9 = 35)

Reusable: Repeat the activity, eliminating the 0 and requiring the addends to be 2- or 3-digit numbers.

Fraction Multiplication

Set up: Write on the board:

1 min = ____ sec   1 hr = ____ min   1 day = ____ hr
1 ft = ____ in.   1 yd = ____ ft
1 cm = ____ mm   1 m = ____ cm   1 km = ____ m

Review these measurement equivalences with students so they can do the problems.

1. **About how long is $\frac{1}{4}$ of a minute in seconds?** (15 sec)
   $\frac{1}{4}$ of an hour in minutes? (15 min)
   $\frac{3}{4}$ of a day in hours? (18 hr)

2. **About how long is $\frac{1}{4}$ of a foot in inches?** (3 in.)
   $\frac{3}{4}$ of a yard in inches? (27 in.)

3. **About how long is $\frac{1}{4}$ of a centimeter in millimeters?** (2 or 3 mm)
   $\frac{1}{4}$ of a meter in centimeters? (25 cm)
   $\frac{1}{4}$ of a kilometer in meters? (250 m)
   $\frac{3}{4}$ of a kilometer in meters? (750 m)

Fraction Multiplication

Set up: Write $\frac{1}{4} * 12 = 3$ on the board.

1. **Write equations to show all the whole numbers that can be produced when multiplying 12 by a unit fraction.** (Equations should yield 1, 2, 3, 4, and 6.)

2. **Write equations to show all the whole numbers that can be produced when multiplying 45 by a unit fraction.** (Equations should yield 1, 3, 5, 9, and 15.)

Whole Number Subtraction

Set up: Write a subtraction problem on the board with blanks in place of the numbers and the digits 0, 1, 2, 3, and 4 to the side of it. Students need slates or paper.

1. **Using all the digits 0 through 4 once, create a subtraction problem with the largest answer possible.** (4,321 – 0 = 4,321)
   When most students have finished the first problem, discuss the strategies students used.
   Now use all the digits 0 through 4 and create a subtraction problem with the smallest whole number answer. (102 – 43 = 59)

2. **Using all of the digits 5 through 9, create a subtraction problem with the largest answer.** (9,876 – 5 = 9,871)
   Now use all of the digits 5 through 9 and create a subtraction problem with the smallest whole number answer. (567 – 98 = 469)

Whole Number Division

Set up: None

1. **How many times can you subtract 6 from 78?** (13)
   How many times can you subtract 12 from 120? (10)
   How many times can you subtract 15 from 180? (12)
   How many times can you subtract 25 from 400? (16)

2. **What are some shortcuts for solving these problems?** (Example: Take away 10 of the number (if possible) first, and then subtract to use up the remainder.)

Operations

Operations

Fraction Addition and Subtraction

Set up: Write the numbers $\frac{1}{2}$, 1, and 2 on the board, leaving space between them. Students need slates or paper.

1. **The answer to $2 - \frac{7}{9}$ is the closest to which of these numbers: $\frac{1}{2}$, 1, or 2?** (1)
   When the class has decided, write $2 - \frac{7}{9}$ under the number 1 on the board. Continue to read the following problems and record each problem under the number closest to the answer.

   $\frac{4}{5} + \frac{5}{6}$ (2) | $1 - \frac{1}{10}$ (1) | $\frac{4}{5} + \frac{1}{6}$ (1)
   $\frac{4}{8} + \frac{1}{9}$ ($\frac{1}{2}$) | $\frac{5}{6} - \frac{1}{3}$ ($\frac{1}{2}$)

2. Discuss strategies that were used in determining solutions. (Example: Decide if each fraction is closer to 0, $\frac{1}{2}$, or 1, and then perform the operation with those estimates.)

3. **Make up some fraction addition problems that are almost equal to 1. Try to find a pattern that will help you.**
   (Examples: $\frac{1}{2} + \frac{2}{5}$, $\frac{1}{2} + \frac{3}{7}$, $\frac{1}{2} + \frac{4}{9}$)

Mixed Operations

Set up: Write the following on the board:
Ticket costs:
Speeding—$10 for each mile over the 55 mph speed limit
Not wearing seat belt—$25

1. **Mr. Smith was a very fast driver and sometimes went over the speed limit. One day a police trooper clocked him driving down a highway at 66 miles per hour and pulled him over. Mr. Smith was also not wearing his seat belt.**
   How much is Mr. Smith likely to pay in fines? ($135)

2. **Mr. Smith told the trooper he was sure he was only going 60 miles per hour. If this were the case, what would his fine be?** ($75)

3. **Could anyone owe $100 for speeding and not wearing a seat belt? Why or why not?** (Example: No. The speeding fine would always be a multiple of 10, ending in a zero, but adding $25 to it would mean the total had a 5 in the ones place.)

Operations

Division Notation

Set up: Write "ten divided by two" on the board. Students need slates or paper.

1. **How many different ways can we write "ten divided by two" using only mathematical symbols?** ($10 \div 2$, $10 / 2$, $\frac{10}{2}$, $2\overline{)10}$)
   Record student responses. Make sure the list is complete by the end of the discussion.

2. **Pick two different division notations and make up a story problem for each.**
   Have the whole class share problems.

Rates

Set up: Draw the chart below on the board. Students need slates or paper.

1. **Michael worked in the yard for 3 hours and made $6. How much did he make for each hour he worked?** ($2) **Complete the chart to show how much Michael would make for 4 or more hours.**
   Discuss results with students.

2. **If Michael got a raise of $0.50 per hour, how would the chart change?**
   Make a third column showing his new pay. ($10, $12.50, $15, $17.50, $20, $50, $75)

Michael's Wages	
Hours Worked	**Pay**
4	($8)
5	($10)
6	($12)
7	($14)
8	($16)
20	($40)
30	($60)

Multiple Operations

Set up: On the board write 5, 2, 1, 0, 5, 7, +, –, ∗, (, and). Below this write 24. Then write 2 ∗ (5 + 7). Students need slates or paper.

1. **Use any combination of the numbers, operations, and grouping symbols on the board to write another name for 24. An example is on the board.**

2. **Now try to use as many of the numbers as you can. Raise your hand if you can use four numbers. Five numbers?**

3. **Did anyone use all six numbers? What is the expression? Does anyone have a different expression using all six numbers?**

Multiple Operations

Set up: On the board write 7, 6, 4, 2, 5, 7, +, −, *, ÷, (, and). Below this write 9. Students need slates or paper.

1. **Use any combination of the numbers, expressions, and grouping symbols on the board to write another name for 9. For example, 4 + 5.**

2. **Now try to use as many of these numbers as you can. Raise your hand if you can use three or more numbers.**
   Write some student expressions on the board.

3. **Did anyone use division? What is the expression? Does anyone have a different expression using division?**

Reusable: Repeat this activity, asking students for six new numbers less than 10 and a new target number.

Multiple Operations: $n * 0 = 0$

Operations

Set up: On the board write 6, 2, 3, 1, 5, 7, +, –, *, (, and). Below this write 0. Students need slates or paper.

1. **Use any of the numbers, operations, and grouping symbols on the board to write another name for 0. For example, (2 * 3) – 6.**
   Tell me your expression so I can record it.
   Does anyone have a longer expression?

2. In the first row of numbers erase the 6 and write 0.
   Now our six is changed to a zero.
   This time I'm going to challenge you to use multiplication and write an expression equal to zero. You can use addition and subtraction also.
   Can you find a longer expression?
   Record student generated expressions on the board. If necessary remind students that zero times any number is zero.

Addition/Subtraction

Set up: Draw the table below on the board. Students need paper or slates and calculators, if available.

Country	Gold	Silver	Bronze	Total
United States	?	39	29	103
China	32	17	14	?

1. **The U.S. won 103 medals in the 2004 summer Olympics. Thirty-nine were silver and 29 were bronze. How many were gold?** (35 medals)

2. **China won 32 gold, 17 silver, and 14 bronze medals. How many total medals did they win?** (63 medals)

3. **A total of 929 medals were given to athletes at the 2004 Olympics. About what fraction of this number did the U.S. win?** ($\frac{103}{929}$ or about $\frac{1}{9}$)

The Median

Set up: The board is needed. Students need slates or paper and calculators, if available.

1. **Determine your age in months and write it down.**
   Let's take a sample of five ages.
   Ask five students to report their ages in months and record them on the board.
   Let's list these in order from greatest to least.
   Record vertically and leave plenty of room between each age for additional data.
   Look at the five pieces of data we have. How do we find the median of the data? (When the data is listed in order, choose the middle piece of data.)

2. **Let's add five more pieces of data.**
   Have five students provide their ages and insert them in the ordered list.
   Now that we have new data, do you think we have a different median? Why?
   How do we find the median of the 10 ages? (With an even number of data points listed in order, choose the number halfway between the "middle" two.)

3. Continue adding ages of different students five at a time until the whole class is represented. Have students guide the recalculation of the median several times, ending with the median of student ages in months for the whole class.

Data

Time Graph

Set up: Sketch two axes on the board. Students need slates or paper.

1. **When Hector was at the park he ran around the track once. He noticed that he could run at about the same speed all the way around. We are going to sketch a graph that shows Hector's speed as he exercised.**
   I'm going to label the horizontal axis "Time."
   What should the vertical axis be labeled? ("Speed")
   Draw and label the horizontal and vertical axes on your paper.

2. **Hector stands on the track waiting to start. Then he takes off.**
   Where should the line start? (at (0,0) or the origin)
   Show me with your hand whether the line stays flat, slants up, or slants down. (Slants up)
   Sketch a small portion of the curve slanting up and then flattening.

3. **Continue the line on your graph to show Hector's speed as he builds up speed, runs once around the track, slows down to a walk and then stops.**

Data

Tally Chart

Set up: Draw a two-column chart on the board. Label the columns "Month" and "Number of Birthdays." Fill in the months of the year.

1. **I need a volunteer to come to the board and tally the Birthday Month data.**
   One at a time call out the month of your birth.
   Be sure that the recorder uses tally marks and does not substitute numbers. If any month has five or more tally marks, make sure to use slant marks to indicate bundles of five.

2. **Which month has the most student birthdays? Which month has the fewest student birthdays? What is the range? What is the mode?**

Mean, Median

Set up: Sketch the line plot below on the board.

Number of Families							
			X			X	
	X	X	X			X	
	0	1	2	3	4	5	6

Children in Family

1. **This line plot shows the number of children in six different families.**
   What is the range? (5)
   What is the median? (2)

2. **What is the mean number of children?** ($2\frac{1}{2}$)
   How did you determine the mean? (Various methods of averaging)

3. **If you think a statement is true, show me a thumbs-up. If you think a statement is not true, show me a thumbs-down.**
   Two families have the mean number of children. (false)
   Two families have more than the mean number of children. (true)
   The median is smaller than the mean because more of the families are small. (true)

Data

Mean

Set up: Write the numbers 10, 4, 5, 2, and 4 on the board.

1. **What is the median of this group of numbers?** (4)
   What is the range? (8)
   What is the mean? (5)

2. **A sixth number is included in the group and the mean goes up.**
   What might that number be? Any other suggestions? (Any number greater than 5)
   A sixth number is included in the original group and the mean goes down.
   Describe the number. (The number is less than 5.)

3. **If you think a statement is true, show me a thumbs-up. If you think a statement is false, show me a thumbs-down.**
   Sometimes the mean is more than the maximum. (false)
   The mean *may* be one of the numbers in the data set. (true)
   The mean *must* be one of the numbers in the data set. (false)

Reading a Bar Graph

Set up: Quickly sketch the bar graph below. Include the numbers along the axes and label the vertical axis.

1. **Sue's mother thought she should help out more. Sue asked some friends how many hours a week they helped their families. She made this graph.**
   How should I label the horizontal axis? (Example: "Hours Worked")
   Where should I write it?
   What is an appropriate title for this graph? (Example: "Number of Hours Children Work") **Where should I write it?**

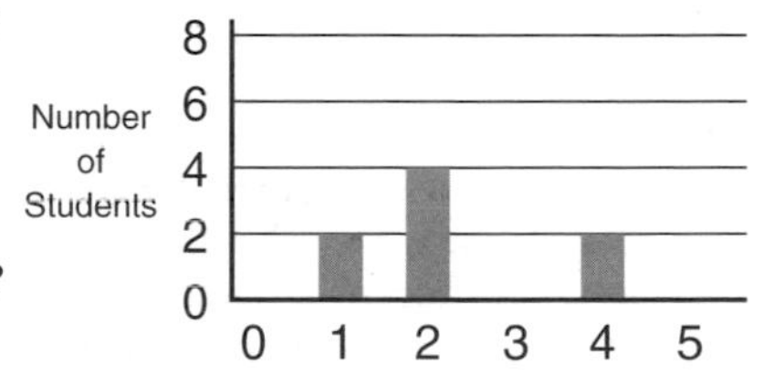

2. **How many friends did Sue ask?** (8)
   How can you find the median of this data? (With an even number of data points listed in order, choose the number halfway between the "middle" two.)
   What is the median? (2)
   Explain how you can find the mean. (Various methods of averaging)
   What is the mean? ($2\frac{1}{4}$)

Data Landmarks

Set up: Draw the chart below on the board. Have each student count the number of pockets in his or her clothes. Record each student's number on the board.

	Median
	Mode
DATA LANDMARKS:	Range
	Maximum
	Minimum

1. **Tell me how to find the median number of pockets a student in this class has. What should I do first?**
   Call on different students for each of the steps. Let the other students make corrections as needed until the median is found and recorded in the chart.

2. **Tell me how to find the mode, the range, the maximum and the minimum.**
   Record each value as it is found.

Reusable: Repeat this activity using another item such as number of books in each desk or number of buttons on clothing.

Mean, Median

Set up: Write the numbers 4, 1, 0, 5, 4, 2, and 0 on the board. Students need slates or paper.

1. **These data represent the number of movies a group of students saw over the vacation.**
   Tell me how to sketch a line plot of this data.

2. **What is the minimum? The maximum?** (0 movies, 5 movies)
   What is the range? (5 movies)
   What is the median? (2 movies)
   How many students saw the median number of movies? (1 student)
   How many saw less than the median number of movies? (3 students)

3. **What is the mean number of movies seen over vacation?** (about 2.3 movies)
   How did you determine the mean? (Various methods of averaging)

Data

Simple Probability

Set up: Draw four circles on the board with the numbers 1, 2, 3, and 4 inside them.

1. **I have these four chips in a bag. Without looking I select one chip. What is the chance of selecting the number 2 chip?** ($\frac{1}{4}$ or equivalent) **The number 3 chip? The number 1 chip? The number 4 chip?** (All $\frac{1}{4}$ or equivalent)
   What do you notice about these chances? (They are equally likely.)
   Have students give examples of equally likely events.

2. **What would be the chance of selecting an even numbered chip? An odd numbered chip?** (Both $\frac{1}{2}$ or equivalent)
   What can you say about these chances? (They have a 50-50 chance.)
   Have students give examples of 50-50 events.

3. **What is the chance that I will select a number 5 chip from the bag?** (0)
   Have students give examples of impossible events.

4. **What is the chance that when I select a chip it will have a 1, 2, 3 or 4 on it?** (1 or 100%)
   Have students give examples of certain events.

Probability: Range

Set up: Draw a horizontal line on the board. Write 0 at the left end, $\frac{1}{2}$ in the middle, and 1 at the right end.

1. **The probability, or chance, of any event is always between 0 and 1. Give me an example of an event that has a probability of 0.** (Example: Rolling a 0 with a six-sided die)
   How about one that has a probability of $\frac{1}{2}$? (Example: Flipping a coin and getting HEADS)
   And one with a probability of 1? (Example: The sun rising)
   Record events under the 0, $\frac{1}{2}$, or 1 as appropriate.

2. **If an event has a probability of 1, what can we say about that event?** (It is certain.)
   If an event has a probability of 0, what can we say about that event? (It is impossible.)

Vocabulary: Experiments, Outcomes

Set up: None

1. **When we talk about probability, we talk about experiments and outcomes.**
   What is an experiment? (An activity with one or more possible results, such as rolling a die)
   What is an outcome? (One of the possible results of an experiment, such as rolling a 2)
   Think of an example of an experiment and an outcome of that experiment.
   Have students share some of their examples.

2. **Let's say our experiment is tossing a coin. What are all the possible outcomes?** (HEADS, TAILS)
   Now let our experiment be rolling a die. What are the possible outcomes for this experiment? (1, 2, 3, 4, 5, 6)
   What if our experiment was picking a card from a deck of cards? What are the possible outcomes? (Each card is an outcome.)

Probability Representation

Set up: Draw a table as shown below (without values).

1. **What are the outcomes for the Toss a Coin experiment?** (HEADS, TAILS)
   What can we say about these two outcomes? Is one more likely than the other? (No, they are equally likely.)

Experiment: Toss a Coin

Outcome	Probability Representations		
H	($\frac{1}{2}$)	(50%)	(0.50)
T	($\frac{1}{2}$)	(50%)	(0.50)

2. **What is the probability that the outcome is HEADS?** ($\frac{1}{2}$ or equivalent)
   Are there other ways to express this probability? (50%, 0.5)
   Discuss fraction, decimal, and percent representations. Fill in the table.

3. **Have you ever heard of probabilities being described in each of these forms? Can you give an example?** (Example: A 50% chance of rain)

Reusable: Repeat this activity with other experiments such as Roll a Die or Pick a Card.

Probability

Doing an Experiment

Set up: Give groups of students a penny. Draw the tally chart below on the board. Students need slates or paper.

Group	H	T
1		
2		
3		
4		
5		
TOTAL		

1. **What is the probability of getting HEADS when I toss this coin? Of getting TAILS?** ($\frac{1}{2}$, $\frac{1}{2}$)

2. **If I toss this penny ten times, how many times should I get HEADS?** (about 5)
   Let's see what happens. Within your group have one person toss the penny 10 times and another person record the results.
   As groups finish, record the results on the board.

3. **Did any of the groups have exactly 5 HEADS and 5 TAILS? Why didn't all of the groups have 5 HEADS and 5 TAILS? Let's total all of the results. Is our total close to half HEADS and half TAILS?**

Probability of Events

Set up: None

1. **Let's pretend to roll a die. What are the possible outcomes?** (1, 2, 3, 4, 5, 6) **What is the probability or chance of rolling a 3?** ($\frac{1}{6}$ or equivalent) **How did you determine this probability?**

2. **An event is something that happens. For example, if you roll a die, rolling an even number is the event that happens when you roll a 2 or a 4 or a 6. How can you find the probability of an event?** Guide the discussion toward the idea that the probability of an event is the number of "favorable" outcomes divided by the total number of outcomes. For example, the probability of rolling an even number is $\frac{3}{6}$ or $\frac{1}{2}$.

3. **What is the probability that we'll roll an odd number when we roll our die?** ($\frac{3}{6}$ or $\frac{1}{2}$) **A number greater than 4?** ($\frac{2}{6}$ or $\frac{1}{3}$) **A number greater than 6?** ($\frac{0}{6}$ or 0)

Probability

Metric Units

Set up: Draw the charts below on the board. Do this activity with the class or have students work in groups and then share.

cm	m

g	kg

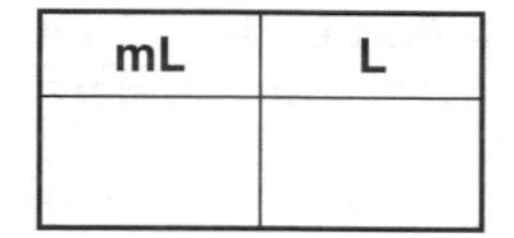

mL	L

1. **Two common units of metric length are the centimeter and the meter. Name some things that are more likely to be measured with centimeters. With meters?** (Examples: Length of a nail in cm; length of a soccer field in m)
   Collect the ideas offered by students on the board.

2. **Now do the same for the common metric units of weight.**

3. **Let's do this again for the common metric units of capacity.**

Acknowledge that some things may be measured in units of both weight and capacity on packaging, as well as in both metric and U.S. customary units.

Metric Length

Set up: A meter stick would be helpful.

1. **Hold up your hands to show about how long a yard is.**
   Now hold up your hands to show about how long a meter is.
   What is the difference? Which is longer? (meter; 1 yard = 36 inches; 1 meter = 39.6 inches)

2. **In the Olympics there used to be an event called the 100-yard dash.**
   Now there is an event called the 100-meter sprint.
   Which run would take longer? (100-meter sprint)
   Do you think there would be a big or a small difference in run times? Why?

3. **The 440-yard run was another old track event. The new event is the 400-meter run.**
   Which run do you think is longer? (They are just about the same.)
   How could you check? (Example: Change both to inches.)

Perimeter

Set up: Write the measure "$8\frac{1}{2}$ inches by 11 inches" on the board.

1. **This describes the measure or size of a normal sheet of paper. What is the perimeter of a sheet of paper?** (39 in.)

2. **Imagine we fold a sheet of paper in half to form a smaller rectangle. What are the possible perimeters of a half sheet of paper?** (28 in., $30\frac{1}{2}$ in.) Review both lengthwise and widthwise folds.

3. **What if we folded the paper in half again to form an even smaller rectangle? What are the possible perimeters of a quarter sheet of paper?** ($22\frac{1}{2}$ in., $26\frac{1}{4}$ in., $19\frac{1}{2}$ in.)

4. **Do you notice any patterns?** (Example: In some cases, the smaller perimeter is 8.5 inches smaller than the perimeter in the previous problem.)

Measurement

Relations: Area of a Triangle

Set up: Students need slates or paper.

1. **Sketch a triangle that has a base of 4 inches and a height of 6 inches. What is the area of the triangle?** (12 in^2)
   Explain how you determined the area.

2. **Sketch another triangle with the same area and a base of 3 inches. Compare the two sketches. What is the height of your new triangle?** (8 in.)

3. **What other lengths for base and height yield a triangle of 12 in^2?** (2 in. × 12 in., 1 in. × 24 in.)

4. **Specify the base and height for different triangles that have an area of 24 square inches.** (1 in. × 48 in., 2 in. × 24 in., 3 in. × 16 in., 4 in. × 12 in., 6 in. × 8 in.)

Volume of a Prism

Set up: Draw a representation of a rectangular prism on the board with empty spaces for its length, width and height labels. Draw a table on the board for length, width and height. Students need slates or paper.

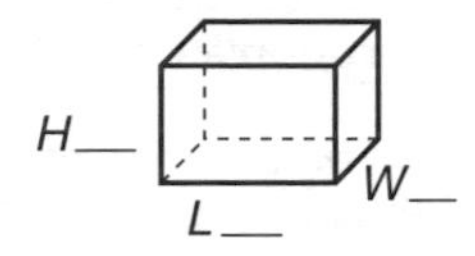

1. **If a rectangular prism has a volume of 36 cubic centimeters, what are some possible dimensions it could have?** (Example: 6 cm × 2 cm × 3 cm) Collect student responses in the table.

2. **Make a table to show the possible dimensions for a rectangular prism with a volume of 54 cubic centimeters.** (Example: 3 cm × 3 cm × 6 cm)

3. **If the length and width of a rectangular prism are both 5 cm, what are some possible volumes it might contain?** (Example: 25 cm^3, 50 cm^3, 75 cm^3)

4. **What patterns do you notice?** (Example: The dimensions are factors of the volume.)

Angle Measure

Set up: On the board, draw and label two analog clock faces, one showing 9:00 and the second showing 11:00. Draw two other circles on the board labeled with the times 9:15 and 9:30, but do not draw the clock hands.

1. **About how many degrees are in the angle formed by the hands at 9:00?** (90°)
   Discuss solutions.

2. **About how many degrees are in the angle formed by the hands at 11:00?** (30°)

3. **How many degrees does the minute hand move in a minute?** (6°) **How did you determine this?**

4. **Draw pictures of clocks for 9:15 and 9:30 and estimate the angle measures that are formed by the hands of each clock.** (a little less than 180° or a little more than 180°, a little more than 90° or a little less than 270°)
   Have students come to the board to show their work.

Scale

Set up: Write “1 inch stands for 1 foot” on the board and sketch a bicycle. Students need slates or paper.

1. **A picture of a bicycle is three inches tall. This scale says that each inch of the picture represents one foot of the real bicycle.**
   Write down how tall the bicycle is. (3 ft)
   How many inches is that? (36 in.)
   How many yards? (1 yd)

2. **Write down your approximate height in feet and inches and in inches. If you used the same scale and sketched a picture of yourself, how tall would the picture be?**

3. **Would the bicycle be a good fit for you? Why or why not?**

Scale Drawings

Set up: Sketch the rectangle below and write "1 unit stands for 2 feet" on the board. Students need slates or paper.

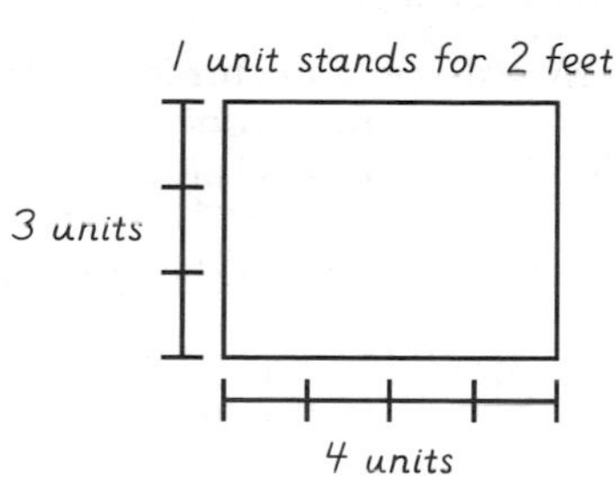

1. **Here is a scale drawing of a garden. Each unit represents 2 feet. How many units long is the drawing?** (4 units)
   How many units wide? (3 units)
   What is the length and width of the actual garden in feet? (8 ft, 6 ft)

2. **What is the perimeter of the actual garden in feet?** (28 ft)

3. **Use the same scale to make a drawing of a garden that is 6 units by 1 unit. What is the perimeter of the garden in feet?** (28 ft)

Measurement

Geometric Properties

Set up: Sketch the figures below on the board. Have students draw the shapes on their slates or paper.

1. **Pick a shape you think doesn't belong with the others. Which one did you pick and why?**
   Put an x through the shape students agree upon.

2. **Pick another shape that doesn't belong with the others. Which one did you pick and why?**
   Put an x through that shape.

3. **There are three shapes left. Is there one that doesn't belong? Why?**
   Put an x through it.

4. **There are only two shapes remaining. What are they? How are these shapes similar? How are they different?**

Geometric Vocabulary

Set up: Write the following prefixes on the board: *tri-, quadra-, penta-, hexa-, hepta-, octo-, nona-, deca-*. Students can record these prefixes on their slates or paper.

1. **These are prefixes that mean different numbers.**
   They are often used in the names of geometric figures.
   As I point to the prefix, tell me the number it means. (3, 4, 5, 6, 7, 8, 9, 10)
   Have the whole class respond.

2. **Now, when I point to a prefix, I want you to hold up the number of fingers the prefix means. For example, if I point to *quadra-*, you hold up 4 fingers. Check your neighbors.**
   Point to the various prefixes in no particular order.

3. **Tell me the names of some shapes that have these prefixes.**
   (Example: triangle)
   For each shape name, have students indicate the number of sides.

4. **Other words also use these prefixes. For example, *tricycle* starts with *tri-* because it has three wheels.**
   Do you know any other words that have these prefixes?

Solid Shapes

Set up: Write the names of some solid shapes on the board, such as cylinder, cone, and square pyramid.

1. **While you are at** __________ (lunch, recess, a field trip, or other activity), **find at least two examples of these solids. Write them down or remember them and we'll talk about what you found later today.**

2. (Later time) **Let's make a list of the solids you saw while at** __________. Record items on the board in shape groups (cylinder, cone, square pyramid).

Reusable: Repeat with other solid shapes such as rectangular prism, triangular prism, sphere, and so on.

Angle Classification

Set up: Draw right, acute, obtuse, and straight angles on the board as below.

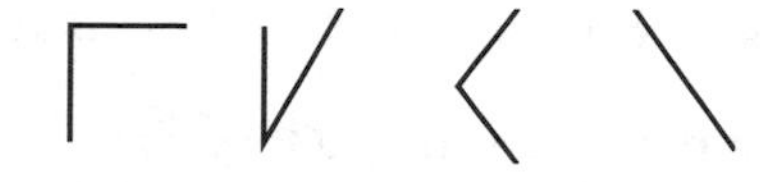

1. **As I point to an angle, tell me its name and how it differs from the other three angles on the board.**

2. **As I say the name of an angle, use your arm and your body to make that type of angle.**
   Randomly call out the angle types and let students look around and check each other's arm angles.

3. **Let's play Simon Says with arm angles.**
   "Simon says right arm acute angle."
   "Simon says left arm straight angle."
   "Simon says left arm right angle."
   "Right arm right angle."
   Did I catch you?

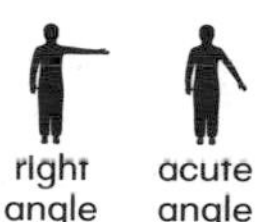

right angle

acute angle

obtuse angle

straight angle

Scale Drawing a Rectangle

Set up: Students need paper and rulers.

1. **A school has a rectangular playground that measures 50 yards by 20 yards.**
   What is the area of the playground? (1,000 yd^2)

2. **Work with a partner to create an accurate scale drawing of the playground.**
   What scale did you use in your drawing?
   Discuss various scales and have students hold up drawings for others to see.

3. **Use the same scale to make a drawing of a different rectangular playground that has the same area as the first one.** (Example: A playground measuring 25 yd by 40 yd)

Geometry

Properties of 2-Dimensional Shapes

Set up: Write the names of the different polygons, 3 to 8 sides, on the board: triangle, quadrangle, pentagon, hexagon, septagon, octagon.

1. **Fold your paper into eight sections. In each of the first six sections, write the name and draw an example of each type of polygon named on the board.**

2. **In the first of the two empty sections, describe all the ways the shapes are the same.** (Example: They all are made up of line segments.)

3. **In the remaining section, explain how they differ. Continue on the back if necessary.** (Example: They have different numbers of sides.)

Symmetry

Set up: Write the numbers 0 through 9 on the board. Students need slates or paper.

1. **Find all lines of symmetry in the numbers zero through nine.** (Example: A vertical line through 8)

2. **Form as many 2-digit numbers as you can that have lines of symmetry.** (Example: 80)

3. **How about 3-digit numbers? Which ones have lines of symmetry?** (Example: 380)

Naming Line Segments

Set up: Draw the chart below and a line segment $\overline{AB}$ on the board. Students need slates or paper.

A B

Number of Points	Names of Segments
2	
3	
4	
5	

1. **Copy the line segment and chart on your paper.**
   Name all the line segments that are made by adding point *C* to line segment $\overline{AB}$, near the middle. ($\overline{AB}$, $\overline{AC}$, $\overline{CB}$)
   Go over the solutions to this part and fill in chart before continuing.

2. **Now put point *D* on the line segment between *A* and *C*.**
   How many different segments can you name now? (6)

3. **Are you beginning to see a pattern? Try adding another point on the line segment to see if your theory works with the new point.**

Geometry

Identifying Characteristics of 2-Dimensional Shapes

Set up: Students need slates or paper.

1. **Draw a figure that has 3 sides but is not a triangle.** (Example: A slice of pizza)
   Have students share their ideas with their group or the class.
   What things must a shape have to be a triangle?
   Create a class definition of a triangle.

2. **Draw a figure that has four angles but is not a quadrilateral.** (Example: The letter *X*)
   Now draw another example that is different.
   Have students share their ideas with their group or the class.
   What properties must a shape have to be a quadrilateral?
   Create a class definition of a quadrilateral.

3. **Draw something that is rounded but is not a circle.** (Example: oval)
   Can you draw another example that is different?
   Have students share their ideas with their group or the class.
   What things must a shape have to be a circle?
   Create a class definition of a circle.

Symmetry: Line of Reflection Outside the Figure

Set up: Draw F and $\overleftrightarrow{AB}$ and $\overleftrightarrow{CD}$ on the board as shown below. Students need slates or paper.

1. **Reflect F in $\overleftrightarrow{AB}$. Draw the reflected image. Describe the reflected image.**

2. **Reflect F in $\overleftrightarrow{CD}$. Draw and describe the reflected image.**

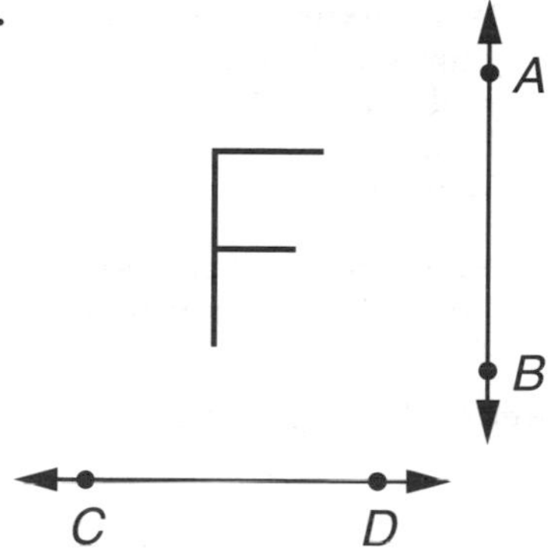

Creating Number Sentences

Set up: Write the following on the board: 3, 4, 5, +, –, =, and 4 + 3 – 5 = 2. Students need slates or paper.

1. **Write expressions for as many different numbers as you can using the numbers 3, 4, and 5 exactly one time each and the symbols *plus*, *minus*, and *equals*. An expression for the number 2 is already done on the board.**
   Write unique solutions on the board.

2. **Now use all of the above symbols except use the multiplication sign instead of the minus sign.** (Example: 3 * 4 + 5 = 17)

Reusable: Repeat this activity with other numbers.

Multiplication Patterns

Set up: Students need slates or paper.

1. **Write the multiples of 4 up to 40. Look through the list and circle multiples of 4 that are also multiples of 8. What is the next number after 40 that is a multiple of both 4 and 8?** (48) **Why do you think this is so?**

2. **Next write the multiples of 6 up to 60. Which of these numbers are also multiples of 9?** (18, 36, 54) **Describe the pattern that you see.**

3. **Look at both lists, the multiples of 4 up to 40 and the multiples of 6 up to 60. What numbers are on both lists?** (12, 24, 36) **What is the next number that is a multiple of both 4 and 6?** (48) **Why do you think this is so?**

Reusable: Repeat this activity with other pairs of numbers such as 2 and 3 or 3 and 9.

Patterns

Set up: Draw the table below on the board.

Number of Dogs	Number of Feet

1. **How many feet are on 3 dogs? 4 dogs? 5 dogs?** (12 feet, 16 feet, 20 feet)
   Write this information in your table.
   What is the general pattern?
   How many feet are on 22 dogs? (88 feet)
   If you know the number of dogs, how can you determine the number of feet?

2. **If there are 16 dog feet in a room, how many dogs are there?** (4 dogs)
   What if there are 48 dog feet in the room? (12 dogs)
   What is the general pattern? If you know the number of dog feet, how can you determine the number of dogs?

3. **Which numbers between 4 and 24 would be unlikely to be the total number of dog feet?** (any number that is not a multiple of 4) **Why?**

Patterns

Set up: Draw the *in/out* table below on the board. Students need slates or paper.

in	*out*
(5)	0
10	5
(26)	21
14	9

1. **Work with a partner. Copy the table and complete it. Fill in three or more blank rows with a partner.**

2. **Use words to write a rule for the *in/out* table.** (Example: Subtract 5.)
   Suppose a mystery number *M* is in the *in* column. How can we determine the output?
   If there is a mystery output *Q*, how can we determine what the input was?

3. **If $8\frac{1}{2}$ is the input, what is the output?** ($3\frac{1}{2}$)
   If 5.2 is the output, what is the input? (10.2)

Reusable: Repeat this activity using decimals or fractions and different operations.

Writing Algebraic Expressions

Set up: Students need slates or paper. Write $3x$ on the board.

1. **Many times we use the letter x to stand for an unknown number. We call x a variable. The $3x$ stands for "3 times x."**
   Here is a story that fits with $3x$: There are 3 snacks in a package, so if I buy x packages, I will have 3 times x snacks.
   Work with a partner and think of another story that fits with $3x$.

2. **Write an algebraic expression for each statement.**
   Terry had x gumballs and then gave away 3. How many does he have now? $(x - 3)$
   A teacher is x years older than her 10-year-old students. How old is the teacher? $(x + 10)$
   Jim is 11. His little sister is x years old. How much older is Jim? $(11 - x)$

3. **We can also use other letters to stand for variables. Let's use n.**
   If n friends each receive one slice of a big pizza, how many slices will we have to cut? (n) **What fraction of the pizza will each friend get?** ($\frac{1}{n}$ of the pizza)

Algebra

Multiplying by $\frac{1}{2}$

Set up: Students need slates or paper. Write $2 * \_\_\_ = 1$ on the board.

1. **Read this sentence. What number will make this statement true?** ($\frac{1}{2}$)
   Draw a picture or tell a story that illustrates $2 * \frac{1}{2} = 1$.
   Hint: Think of the relationship between 1 and 2. One is ______ of 2.
   (half)
   If we divide instead of multiply, what number will go in the blank?
   Write $2 \div \_\_\_\_ = 1$. (2)

2. **Work with a partner and try these problems.**
   $12 * \_\_\_\_\_ = 6$. ($\frac{1}{2}$)
   $12 \div \_\_\_\_\_ = 6$. (2)
   $10 * \frac{1}{2} = \_\_\_\_\_$. (5)
   $10 \div 2 = \_\_\_\_\_$. (5)

3. **Finish this sentence: Any number multiplied by one-half is the same as ___________.** (that number divided by 2)

Algebra

Whole Number Sums and Differences

Set up: Sketch a two-column table on the board. Students need slates or paper.

1. **Tell me a pair of whole numbers whose sum is 5.** (Example: 1 and 4)
   What other pairs of whole numbers have a sum of 5?
   Fill in the chart as pairs are suggested.
   As the first number of a pair increases, what happens to the second number? (It decreases.)

2. Draw a new two-column chart.
   Tell me a pair of whole numbers whose difference is five. (Example: 7 and 2)
   Record the pairs in the chart.
   Work with a partner and find several pairs of whole numbers whose difference is 5.
   Look at the pairs you have found. As the first number of a pair increases, what happens to the second number? (It increases.)

3. **How many pairs of whole numbers have a sum of 5?** (3) **Explain.**
   How many pairs of whole numbers do you think have a difference of 5? (an unlimited number) **Explain.**

Formula

Set up: Write $F = 3Y$ on the board.

1. **This formula can be used to find how many feet there are in any number of yards.**
   How many feet are in 4 yards? (12 ft) **10 yards?** (30 ft) **2 yards?** (6 ft)

2. **How can you find out how many yards there are in any number of feet?**
   How many yards are in 9 feet? (3 yd) **15 feet?** (5 yd)
   How can you write the formula? ($Y = \frac{F}{3}$)

3. Sketch this two-column table on the board.
   Tell me the missing values in this table.

Feet	Yards
6	(2)
(21)	7
15	(5)
(12)	4
5	($\frac{5}{3}$ or $1\frac{2}{3}$)

Algebra

Sequences

Set up: Write ___, 4, 7, 10, 13 on the board. Students need slates or paper.

1. **Sandra thought of a rule and then wrote this sequence of numbers. What do you think her rule was?** (Add 3) **Explain.**

2. **What number comes after 13 in the sequence?** (16) **What number comes before 4?** (1) **Explain.**

3. **Work with a partner and determine if 19 is a number in the sequence.** (yes) **How did you decide?**

Reusable: Repeat this activity with any sequence and missing numbers in the sequence.

Sequences

Set up: Write 1, $1\frac{1}{2}$, 2, $2\frac{1}{2}$, . . . on the board. Students need slates or paper.

1. **The rule for this sequence is "Add $\frac{1}{2}$ to the previous number." What are the next three numbers in this sequence?** (3, $3\frac{1}{2}$, 4)
   Add these numbers to the sequence on the board.

2. **Let's write the same sequence of numbers using fractions in place of mixed numbers. We can begin with 1, $\frac{3}{2}$, 2. What would be next?** ($\frac{5}{2}$, 3, $\frac{7}{2}$, 4)
   Record these fractions under the sequence.

3. **Work with a partner and determine if 15 is a number in the sequence.** (yes) **How did you decide?**

Algebra

Sequences

Set up: Write $\frac{1}{3}$, $\frac{2}{3}$, 1, $\frac{4}{3}$, . . . on the board. Students need slates or paper.

1. **Alicyn thought of a rule and then wrote this sequence of numbers. What do you think her rule was?** (Add $\frac{1}{3}$) **Explain.**

2. **What is the next number in the sequence?** ($\frac{5}{3}$)
   Write down the next three numbers of the sequence. ($\frac{5}{3}$, 2, $\frac{7}{3}$)

3. **Work with a partner and determine if 3 is a number in the sequence.** (yes) **How did you decide?**

Grouping Symbols

Set up: Write the problem groupings from #1 and #2 below on the board. Students need slates or paper.

1. **Let's solve these problems together.**

   $3 * (2 + 6) = (24)$ $5 - (2 + 3) = (0)$

   $(3 * 2) + 6 = (12)$ $(5 - 2) + 3 = (6)$

   You can see that even though the problems in each pair look similar, they have different answers. Why is this?

2. **Now solve this set of problems. When you are finished, check your answers with a partner.**

   $(8 \div 2) + 2 = (6)$ $12 - (6 \div 3) = (10)$

   $8 \div (2 + 2) = (2)$ $(12 - 6) \div 3 = (2)$

Using Parentheses

Set up: Write the problem groupings from #1 and #2 below on the board. Students need slates or paper.

1. **Let's do these problems together.**
   $7 + 4 \div 2 = (9)$
   $(7 + 4) \div 2 = (\frac{11}{2} \text{ or } 5\frac{1}{2})$
   $7 + (4 \div 2) = (9)$
   What causes the answers to be different?
   What rules do we follow when using parentheses?

2. **Copy the first problem in this set. Add parentheses and solve it.**
   Now put parentheses in a different place and solve.
   Compare what you did with your neighbor.
   Continue with the other problems.
   $3 - 1 \div 2 =$ $\qquad$ $8 \div 4 \div 4 =$
   $16 + 4 \div 2 =$ $\qquad$ $9 - 6 - 3 =$

3. **What interesting things did you notice about these problems?**

Algebra

Fractions

Set up: Write $\frac{3}{2}$ on the board.

1. **Tell me a fraction that is equivalent to $\frac{3}{2}$.** (Example: $\frac{6}{4}$) Write the fraction on the board. **Another one?** Repeat until you have written about 20 fractions equivalent to $\frac{3}{2}$.

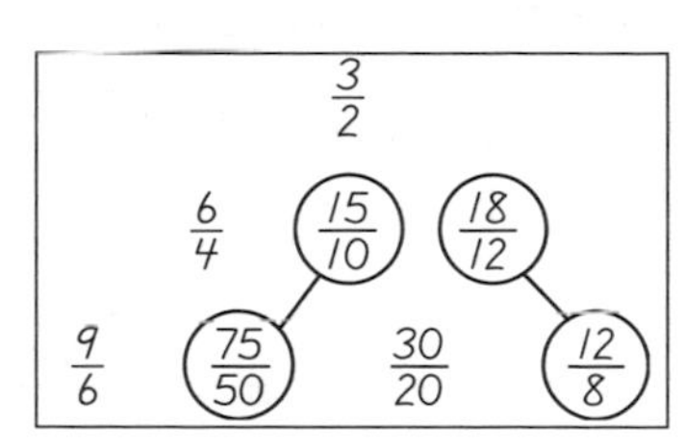

2. **Come to the board. Loop together two fractions that have a sum of 3.** Continue to invite students to loop together pairs of fractions until the class concludes that any two fractions on this list add up to three.

3. **Now let's use a different color of chalk to loop fractions that add up to $4\frac{1}{2}$.** (Any 3 fractions)

Reusable: Repeat this activity using fractions equivalent to $\frac{4}{3}$.

Numeration

Powers of Ten

Set up: Start a three-column chart and write 10^3 in the fourth row, first column.

(Power of 10)	(Standard Notation)	(Multiply by 10s)
10^3	(1,000)	(10∗10∗10)

1. **This number is a power of 10. How is this number read?** (Example: "ten cubed")
   Tell me how to write "ten cubed" or "ten to the third power" in standard notation. (1,000)
   How would we multiply by 10s to get this number? (10∗10∗10)
   Write responses in the second and third columns.
   What is the 3 in "ten to the third" called? What is the 10 called? (exponent, base)

continued

2. Write 1,000,000 in the fifth row, second column of your chart.
   Don't call out, but show me by holding up fingers: What exponent is used to write one million as a power of 10? (6)
   Survey student answers and write 10^6 in the first column of your chart.
   How is this read? ("ten to the sixth power")
   How should we write one million as a product of 10s?
   (10*10*10*10*10*10)
   Complete the fifth row of the chart.

3. **How should I label each of these columns in the chart?**
   Now hold up fingers to show what exponent is used to write one billion. (9) **Tell me how to write one billion in standard notation and as a product of 10s.** (1,000,000,000; 10*10*10*10*10*10*10*10*10)
   Complete another row of the chart using one billion.

4. **What patterns do you notice in the chart?**

5. **Using the patterns you found, complete the first three rows for 10^0, 10^1, and 10^2.**

Numeration

Roman Numerals: I–LII

Set up: Draw the chart below on the board.

1. **Roman Numerals are an old system of writing numbers. You sometimes see them on monuments or buildings. What do you think I equals? What about V? And X?** (1, 5, 10) **Tell me how to fill in the blanks.** (XXI, XXII, XXIII, XXV, XXVII, XXVIII)

Roman Numerals	Standard Notation
XX	20
(XXI)	21
(XXII)	22
(XXIII)	23
XXIV	24
(XXV)	25
XXVI	26
(XXVII)	27
(XXVIII)	28
XXIX	29

continued

2. **Look at 29. How do you think 30 is written in Roman Numerals? 31? 39?** (XXX, XXXI, XXXIX)
   What pattern do you notice about the placement of I with another symbol?
   Tell me how to fill in 30 through 39. (XXX, XXXI, ..., XXXIX)
   What patterns do you see in the Roman Numerals for 20 through 29 and those representing 30 through 39?

3. **Fifty is written as L. Tell me how to write 49 and 51.** (XLIX, LI)
   Let's complete the column of Roman Numerals to 52. (XL, XLI, . . ., LII)
   Describe the patterns that you see.

Numeration

Roman Numerals: I–L (Review)

Set up: The board is needed. Students need slates or paper.

1. **Quickly write down the Roman Numerals 1 through 10.** (I, II, . . ., X)
   How can you easily change your list to the Roman Numerals for 11 through 20? How about 21 through 30? (Example: Add an X to the left of each numeral.)

2. **What symbols are needed to write the Roman Numerals for numbers 1 through 30?** (X, V, I)
   What additional symbol is needed to continue to 40? (L)
   Start with 8 and skip count by 10 to 48. Write this sequence of numbers in Roman Numerals. (VIII, XVIII, XXVIII, XXXVIII, XLVIII)

3. **Is the following statement true or false?**
   As numbers get larger, it takes more symbols to write the number using Roman Numerals. (false, or not always true)
   Think of three examples to support your answer.

Scientific Notation

Set up: The board is needed.

1. **The diameter of the Sun is thought to be 870 thousand miles. Tell me how to write 870 thousand in standard and scientific notation.** Write these two representations on the board. (870,000; $8.7*10^5$)

2. Write $1.4*10^6$ kilometers on the board. **This is the measurement of the Sun's diameter in kilometers. How is this number read?** ("one point four times ten to the sixth power") **How can it be written in standard notation?** (1,400,000)

Reusable: Repeat this activity using data from the following chart.

Planet	Diameter (mi)	Diameter (km)	Distance from Sun (mi)	Distance from Sun (km)
Mercury	3,032	4,880	36,000,000	57,900,000
Jupiter	88,736	142,800	483,880,000	778,300,000
Neptune	30,775	49,528	2,796,460,000	4,497,000,000

Source: Beth Rowen and Curtis Slepian (Eds.), *Time for Kids Almanac 2005* (Needham, Mass.: Pearson Education, 2004), 216–219.

Divisibility Rules for 3

Set up: None

1. **If we divide a number by 3 and the remainder is zero, then the number is said to be divisible by 3.**
   Is the number of students in the class divisible by 3?
   Is the number of feet in the class plus the number of heads divisible by 3? (yes)
   How about the number of days in a week? (no) **The number of days in this month? Is the number of months in a year divisible by 3?** (yes)

2. **Here is a shortcut for knowing whether a number is divisible by 3 without doing any dividing: Add up the digits in the number. If the sum of the digits is divisible by 3, then the number itself is divisible by 3. Let's try the shortcut to determine whether the following numbers are divisible by 3. If it is, show me with a thumbs-up. If it isn't, show me with a thumbs-down.**
   24, 42, 50, 57, 100, 150 (yes, yes, no, yes, no, yes)

3. **Explain how you can tell whether the number of days in this year is divisible by 3.**

Fractions: $\frac{1}{2}$ and $\frac{1}{4}$

Set up: The board is needed. Students need slates or paper.

1. **If 12 students are sitting at one table and half of them have milk, how many have milk?** (6 students)
   If fewer than half of those with milk have chocolate milk, how many could have chocolate milk? (0, 1, or 2 students)
   Four of the 12 students have pizza. What fraction of the group has pizza? Is this fraction less than half or more than half? Explain. ($\frac{1}{3}$, less)

2. On the board write down the names of a group of four students in which one student is wearing sneakers.
   How many students out of the total group are wearing sneakers? (1)
   Let's add some students to our group, but be sure that one-quarter of the group is still wearing sneakers.
   Describe our new group by saying the number of students wearing sneakers out of the total. Write the corresponding fraction. (Example: $\frac{3}{12}$)
   What other size groups can be created where one-fourth of the group is wearing sneakers? (Examples: 4, 8, 12) **What fraction of students in each group is not wearing sneakers?** (Examples: ($\frac{3}{4}$, $\frac{6}{8}$, $\frac{9}{12}$)

Numeration

Fractions: <1, = 1, >1

Set up: Write 3 and 9 on the board. Students need slates or paper.

1. **Using only these numbers, write as many different fractions as you can.** After a minute write them on the board. ($\frac{3}{9}$, $\frac{9}{3}$, $\frac{3}{3}$, $\frac{9}{9}$)
   Tell me how to order the fractions from smallest to largest. ($\frac{3}{9}$, $\frac{3}{3}$ and $\frac{9}{9}$, $\frac{9}{3}$)
   For each fraction give a unit fraction or whole number that is equivalent. ($\frac{1}{3}$, 1, 1, 3)

2. **Now expand your list of fractions by writing all the fractions you can using 3, 9, and 12. Sort the fractions into three groups: less than 1, equal to 1, and greater than 1.** ($\frac{3}{9}$, $\frac{3}{12}$, and $\frac{9}{12}$; $\frac{3}{3}$, $\frac{9}{9}$, and $\frac{12}{12}$; $\frac{9}{3}$, $\frac{12}{3}$, and $\frac{12}{9}$)
   Which groups of fractions are improper fractions? (equal to 1, greater than 1)
   Write the fractions in order. Which is the smallest? Next? . . . ($\frac{3}{12}$, $\frac{3}{9}$, . . .)
   Write each fraction in the ordered list in simplest form. What do you notice about the two ordered lists?

Reusable: Repeat this activity with other sets of numbers such as 2, 4, and 8; or 4, 8, and 20.

Decimal Place Value

Set up: Draw the first chart on the board. While students are working on the first chart, you can draw the second chart on the board. Students need slates or paper.

1. **Copy this chart and fill in the missing amounts.**

	The Number 0.1 More	The Number 100 Times More
1.20	(1.30)	(120)
201.00	(201.10)	(20,100)
77.010	(77.110)	(7,701.0)

2. **Now let's complete this one.**

	The Number 10 Times Less	The Number 100 Times Less
50.37	(5.037)	(0.5037)
29.344	(2.9344)	(0.29344)
6.04	(0.604)	(0.0604)

Whole Number Factors

Set up: Write the numbers 36, 100, 144, and 1,000 on the board.

1. **What pairs of numbers, or factor pairs, multiply to make 24?** (1 and 24, 2 and 12, 3 and 8, 4 and 6)

2. **What factor triples multiply to make 24?** (Example: 2, 3, and 4)
   You may want to check solutions to the first two problems before assigning numbers 3 through 5.

3. **Using any whole number except the number 1, find the longest string of factors that multiply to make 24.** (2, 2, 2, and 3)

4. **What are the longest factor strings that multiply to make 36, 100, 144, and 1,000?** (2, 2, 3, and 3; 2, 2, 5, and 5; 2, 2, 2, 2, 3, and 3; 2, 2, 2, 5, 5, and 5)

5. **What do you notice about the numbers in the strings?** (Example: They are all prime numbers.)

Whole and Decimal Number Rounding

Set up: Draw the first chart on the board. While students are working on the first chart, you can draw the second chart on the board. Students need slates or paper.

1. **Copy this chart and round the numbers to the places given.**

	Nearest 100,000	Nearest 1,000	Nearest 100
61,089	(100,000)	(61,000)	(61,100)
362,557	(400,000)	(363,000)	(362,600)
9,999,999	(10,000,000)	(10,000,000)	(10,000,000)

2. **Now let's do some decimal rounding.**

	Nearest Tenth	Nearest Hundredth	Nearest Thousandth
0.467	(0.5)	(0.47)	(0.467)
23.095	(23.1)	(23.10)	(23.095)
100.9999	(101.0)	(101.00)	(101.000)

Numeration

Whole Number Rounding

Set up: Draw the chart below on the board. Students need slates or paper.

Nearest Ten Is 10	Nearest Hundred Is 100	Nearest Thousand Is 1,000	Nearest Ten-Thousand Is 10,000
11		1,358	

1. **Copy this chart. In each box list at least five numbers that would be rounded to the number shown. I've done two examples.**
   Review some student solutions and write them in the chart.

2. **Let's add a row below the chart and write the range of whole numbers that would be rounded to the number shown.** (5–14; 50–149; 500–1,499; 5,000–14,999)

Fraction/Decimal/Percent Conversion

Set up: Draw the chart below on the board.

1. Let's change the fractions in the chart to decimals and percent.

Fraction	Decimal	Percent
$\frac{1}{2}$	(0.5)	(50%)
$\frac{2}{2}$	(1)	(100%)
$\frac{3}{2}$	(1.5)	(150%)

Review each part with students as you complete the chart.

2. Let's do the same with the fractions $\frac{1}{4}$, $\frac{2}{4}$, $\frac{3}{4}$, $\frac{4}{4}$, and $\frac{5}{4}$.

3. Using the same pattern, let's do the same with eighths.

4. Now let's try thirds and sixths.
Non-terminating decimals will need to be discussed as they come up.

Numeration

Decimal Place Value

Set up: Write the following numbers on the board in a column: 12,345; 1,234.5; 123.45; 12.345; 1.2345.

1. **Use expanded notation to show the place value of each digit in the numbers shown on the board. Let's do one together.** (Example: $1{,}234.5 = 1{,}000 + 200 + 30 + 4 + \frac{5}{10}$)

2. **Write the name for each of the numbers in words. Let's do one together.** (Example: 1,234.5 is written "one thousand two hundred thirty-four and five tenths.")

Whole Number Estimation, Multiplication, Division, Measurement

Set up: None

1. **If a student book bag weighs about 15 pounds, how many total pounds of books are the students in this class carrying?** (number of students * 15)

2. **Approximately how many students are in our school?** (number of classrooms * average number of students)
   Estimate the total number of pounds of books all of the students carry in one day.
   Do you think this is an overestimate or an underestimate?

3. **How many pounds are in a ton?** (2,000)
   Now let's estimate how many tons of books the students in the school carry in one day. (answer from #2 ÷ 2,000)

Reusable: Repeat this activity with the number of minutes each student spends on schoolwork each night. How many total minutes do all the students spend on schoolwork? How many hours of study time is this each night?

Operations

Whole Number Division

Set up: The board is needed. Students need slates or paper.

1. **Using a number greater than 20, make up a division problem with a divisor of 7 and a remainder of 2.** (Example: 23 / 7 ⟶ 3 R2)
   List some responses on the board.

2. **Make up two different word problems that go with your division problem.**
   Have students share problems.

3. **List all the numbers less than 70 that can be used to make up a problem with a divisor of 7 and a remainder of 2.** (2, 9, 16, 23, 30, 37, 44, 51, 58, 65)
   Do you notice a pattern? (Example: All numbers are 2 more than a multiple of 7.)

Reusable: Ask students to make up division problems with a divisor of 4 and remainder of 3 or a divisor of 9 and remainder of 4.

Whole Number Division

Set up: Students need slates or paper.

1. **Create division problems that have a divisor of 24 and $\frac{1}{2}$ in the quotient.** (Example: 60 / 24 = $2\frac{1}{2}$)
   What pattern can you use to help find such problems? (Example: The dividend must be 12 more than a multiple of 24.)

2. **List all the dividends less than 200 that work for this activity.** (12, 36, 60, 84, 108, 132, 156, 180)

3. **What would the dividends be if we wanted $\frac{1}{3}$ in the quotient?** (8, 32, 56, 80, 104, 128, 152, 176, 200)

Fraction Addition

Set up: Write $\frac{1}{8} + \frac{3}{8} = \frac{1}{2}$ on the board. Students need slates or paper.

1. **Make up six different fraction addition problems that have a sum of $\frac{1}{2}$. One example is on the board.**
   Explain a pattern that could be used to create such problems.
   (Example: Add a unit fraction to a fraction with the same denominator and a numerator equal to 1 less than half that denominator.)

2. **Make up six different fraction subtraction problems that have a difference of $\frac{1}{2}$.** (Example: $\frac{3}{4} - \frac{1}{4} = \frac{1}{2}$)
   Explain a pattern that could be used to create such problems.
   (Example: Use a fraction with the numerator 1 more than half the denominator minus the unit fraction with the same denominator.)

Fraction Addition

Set up: Students need slates or paper and calculators, if available.

1. **Estimate the sum of $\frac{1}{2} + \frac{1}{3} + \frac{1}{4}$.** (about 1)
   Find the exact answer and change it to a decimal. ($\frac{13}{12}$, ≈ 1.083)

2. **Estimate the sum of $\frac{1}{5} + \frac{1}{6} + \frac{1}{7}$.** (about $\frac{1}{2}$)
   Find the exact answer and change it to a decimal. ($\frac{107}{210}$, ≈ 0.509)
   Go over the solutions to these two before having students proceed.

3. **Estimate the sum of $\frac{1}{8} + \frac{1}{9} + \frac{1}{10}$.** (about $\frac{1}{3}$)
   Find the exact answer and change it to a decimal. ($\frac{121}{360}$, ≈ 0.336)

4. **What pattern do you notice?** (Example: Sums are getting smaller.)
   What do you think a good estimate for the sum of the next three unit fractions would be? ($\frac{1}{4}$)

Integer Addition and Subtraction

Set up: Write the numbers 0 through 9 and 10 through 19 on the board. Students need slates or paper.

1. **List the digits from 0 to 9 on your slate. Make the even numbers negative. Now add all the numbers.** (5)

2. **What happens if you do the same thing for the numbers 10 through 19?** (The sum is 5.)

3. **Will the same thing happen with the next 10 numbers, 20 through 29?** (yes)

4. **Why did both results happen?** (There are 5 even numbers and 5 odd numbers, but each odd number is 1 more than the even number that precedes it.)

Fraction Subtraction

Set up: Write the following subtraction problems on the board:

$\frac{1}{2} - \frac{1}{4}$

$\frac{3}{4} - \frac{1}{8}$

$\frac{1}{2} - \frac{1}{3}$

$\frac{2}{3} - \frac{1}{2}$

Students need slates or paper.

1. **How do you think we should do the first example?**
   Write student solution ideas on the board. ($\frac{1}{4}$)

2. **Now complete the problems.** ($\frac{5}{8}$, $\frac{1}{6}$, $\frac{1}{6}$)
   Have selected students show their work on the board.

Reusable: Write the following on the board for more challenge:

$1\frac{1}{6} - \frac{2}{3}$ ($\frac{1}{2}$) $1\frac{1}{2} - \frac{3}{4}$ ($\frac{3}{4}$) $1\frac{1}{8} - \frac{3}{4}$ ($\frac{3}{8}$)

Operations

Operations

Decimal Addition

Set up: Write the following on the board:

$10 = $8.44 + 1¢ + 5¢ + 25¢ + 25¢ + $1

$6.25

$4.18

Students have play money (optional).

1. **You go to the drug store and your bill is $8.44. If you give the cashier a $10 bill, she may count back the change like this: 1 cent is 45, and 5 cents is 50, and 25 cents is 75, and 25 cents is 9 dollars, and 1 dollar is 10 dollars. She gives you $1.56 in change.**

   Now let's say the bill is $6.25 and you are the cashier. Draw pictures to show how you would count back the change from $10. (75 cents is 7 dollars, and 3 dollars is 10 dollars. The change is $3.75.)
   Review the proper change with the entire class.

2. **The bill is now $4.18. What is the change?** (2 cents is 20, and 5 cents is 25, and 75 cents is 5 dollars, and 5 dollars is 10 dollars. The change is $5.82.)

Percent

Set up: Write the following amounts on the board:

DVD Player	$179
TV	$289
Walkman®	$89
Boom Box	$129

Students need slates or paper.

1. **What would be the discount for each item during a 20% off sale?** ($35.80, $57.80, $17.80, $25.80)

2. **Is there a quick way to find 20% of an amount?** (Example: Determine a 10% discount and double it.)

3. Write the following amounts on the board:
   $15.00 $25.00 $32.50 $56.83

   For very good service, a restaurant patron will leave a 20% tip. If this were the case, what would someone pay *in total* for each meal price shown on the board? ($18.00, $30.00, $39.00, $68.20)

Decimal Addition

Set up: Write the following amounts on the board:

0.4	0.25
0.04	0.025
0.004	0.0025

Students need slates or paper.

1. **What amount must you add to 0.4 to get 1.0?** (0.6)
   What amount must you add to 0.04 to get 1.0? (0.96)
   How did you determine your answer?
   What amount should we add to 0.004 to get 1.0? (0.996)

2. **Now find the amounts we would need to add to 0.25, 0.025 and 0.0025 to get 1.0.** (0.75, 0.975, 0.9975)
   Review results and methods used.

Rates

Set up: Write "Speed Limit: 40 mph" on the board.

1. **Mrs. Johnson needed to buy meat for dinner, but the supermarket was closing in 5 minutes. She lived 2 miles from the market. Driving at the speed limit, could she get there before it closed?** (yes) **If so, how many minutes did she have to shop? And, if not, how many minutes was she late?** (She had 2 minutes to shop.)

2. **She drove home at a more leisurely speed of 30 mph. How long did it take her to get home?** (4 minutes)

Source: Kevin Seabrooke (Ed.), *The World Almanac for Kids 2005* (New York: World Almanac Education Group, 2004), 26.

Operations

Order of Operations

Set up: On the board write 1, 2, 3, 4, 6, 9 and above this write 5. Then write $2^3 - 4 + 1$. Students need slates or paper.

1. **Use any of these numbers and addition, subtraction, multiplication, division, parentheses, and/or exponents once to write another name for 5. An example is on the board.**
   Raise your hand if you can use four numbers. Five numbers?

2. **Did anyone use all six numbers?**
   What is the expression? (Examples: $3 * 2 - 6 + 9 - 4 * 1$, $6^2 - (3 * 9) - 4 \div 1$)
   Does anyone have a different one?

Multiple Operations

Set up: Write 1, 2, 3, 4, 5, and 9 and $+$, $-$, $*$, $\div$, and $\sqrt{\ }$ on the board. Above the numbers write a 2. Students need slates or paper.

1. **Use as many of these numbers and operations as you can to write another name for 2. For example, $4 \div 2$.**
   Raise your hand if you can use four numbers.
   What is the expression? (Example: $9 - 5 - 2 * 1$)
   Are there others?

2. **Try to write an expression that uses the square root operation. Use as many numbers as you can.** (Example: $\sqrt{9} + 5 - 4 - 2$)

Reusable: Ask students for six numbers less than 10 and a target number.

Operations

Operations

Square Root Sign

Set up: Write $\sqrt{9}$ and draw a 3-inch square on the board. Students need slates or paper.

1. **How is this read?** ("the square root of nine")
   What does "square root of 9" mean? (Examples: The number multiplied by itself that equals 9. The side length of a square with an area of 9.)

2. **What number has a square root of 2?** (4)
   Using the square root sign, write an expression equal to 2. ($\sqrt{4}$)
   Now use the square root sign and write another name for 10 and another name for 1. ($\sqrt{100}$, $\sqrt{1}$)

3. **The area of a square is 36 square inches. Write two different names for the length of each side.** (6 in., $\sqrt{36}$ in.)

4. **The area of a square is 8 square feet. Write the length of each side.** ($\sqrt{8}$ in.)

Rates

Set up: Draw the chart below on the board. Students need slates or paper.

Distance

	1 hour	2 hours	3 hours	5 hours	10 hours
Car A: 40 mph	(40 mi)	(80 mi)	(120 mi)	(200 mi)	(400 mi)
Car B: 60 mph	(60 mi)	(120 mi)	(180 mi)	(300 mi)	(600 mi)

1. Two cars left from the same location and began traveling at the same time and in the same direction. One car drove at 40 mph and the other car drove at 60 mph. Copy and complete the chart on the board.

2. Make a chart showing the distance between the two cars after 1, 2, 3, 5, and 10 hours.

Distance Apart	(20 mi)	(40 mi)	(60 mi)	(100 mi)	(200 mi)

3. After one hour the cars are only 20 miles apart, but after 10 hours they are 200 miles apart. How is this possible?

Ratios

Set up: Students need slates or paper.

1. **In an adult, the ratio of the head length to the body length is about 1 to 8.**
   How can we write this ratio using mathematical symbols? (Examples: 1:8, $\frac{1}{8}$)

2. **I am about _______ inches tall. About how many inches would you expect the length of my head to be?**

3. **The ratio of head length to body length in children is different and changes with age. A newborn has a head to body ratio of about 1 to 4. How should we write this?** (Examples: 1:4, $\frac{1}{4}$)
   If a typical newborn is about 21 inches long, about how long is its head? (about 5 in.)

Proportions

Set up: The board is needed. Students need slates or paper.

1. **If a large pizza will feed 4 people, how many pizzas would we need to feed 18 people?** (5 pizzas)

2. **How many pizzas would we need to feed 30 people?** (8 pizzas)

3. Write the following proportions on the board.

   $\frac{1}{4} = \frac{x}{18}$

   $\frac{1}{4} = \frac{x}{30}$

   How can we use these proportions to solve our pizza problems?

Fractions

Set up: Students need slates or paper.

1. **The U.S. won 103 medals in the 2004 Summer Olympics. The total won by all countries was 929.**
   About what fraction of the total medals did the U.S. win? (about $\frac{1}{9}$)

2. **Russia received 92 medals and China won 63.**
   Of the total 929 medals, about what fraction did the U.S., Russia, and China win all together? (about $\frac{1}{3}$ or $\frac{3}{10}$)

3. **About what fraction of the medals were won by Russia?** ($\frac{1}{10}$)
   By China? ($\frac{1}{15}$)

Fraction Addition

Set up: Write the fraction problems shown below on the board:

$\frac{1}{2} + \frac{1}{2}$

$\frac{1}{2} + \frac{1}{3}$

$\frac{1}{2} + \frac{1}{4}$

$\frac{1}{2} + \frac{1}{5}$

Students need slates or paper.

1. **When adding fractions you can always find a common denominator by multiplying the denominators. Use this approach to solve the problems on the board.** ($\frac{4}{4}$, $\frac{5}{6}$, $\frac{6}{8}$, $\frac{7}{10}$)

2. **What pattern or patterns do you notice?** (Example: Numerators increase by 1 as denominators increase by 2.)
   Try a few more to see if the pattern continues.

Reusable: Repeat this activity using $\frac{1}{3} + \frac{1}{3}$, $\frac{1}{3} + \frac{1}{4}$, $\frac{1}{3} + \frac{1}{5}$, and $\frac{1}{3} + \frac{1}{6}$ to see what patterns develop.

Graphs: Actions over Time

Set up: Draw a sketch of a Ferris wheel ride as in the illustration below. Next to this draw a set of coordinate axes. Label the vertical axis "Distance from the Ground" and the horizontal axis "Time." Encourage students to use their slates or paper and draw graphs.

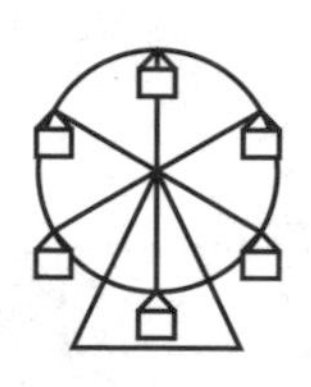

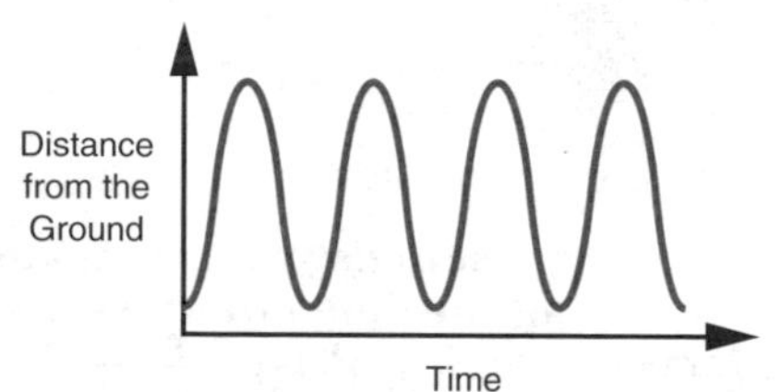

1. **Notice these axes are labeled "Time" and "Distance from the Ground." Let's draw a graph of a person riding a Ferris wheel. Our graph should show how the person moves toward and away from the ground as time passes. How should we start?**
   Draw exactly what students tell you and let the group correct errors until the graph looks similar to the illustration.

continued

2. **Now let's draw a graph of students playing basketball.**
   Draw a clean set of axes on the board. Make sure you leave plenty of room to the right.
   We will graph the distance of the ball from the floor as time passes.
   How shall I label the axes? (Example: "Time" and "Distance from Floor")
   Describe how to draw the following: A player bounces the ball 3 times and throws it to another player who throws it in the basket.

3. **Let's continue on the same graph.**
   The ball bounces when it leaves the basket and the player who gets it dribbles it 8 times.
   This player then throws it to a different player who bounces it 5 times and passes it to a teammate.

Reusable: Repeat this activity using different sports. Let students suggest sports and describe various scenarios.

Data

Data Landmarks

Set up: Have each student count the number of pencils in his or her desk. Record each student's number on the board.

1. Write the word *median* on the board.
   Tell me how to find the median number of pencils students in this room have.
   Call on different students for each of the steps. Let the other students make corrections as needed until the median is found and recorded.

2. Write the words *mode*, *range*, *maximum*, and *minimum* on the board.
   Tell me how to find the mode, the range, the maximum, and the minimum.
   Record their values as they are found.
   How should I find the mean?
   Add the word *mean* to your list. Look for outliers and discuss what measure of central tendency gives the most typical number of pencils per student.

Reusable: Repeat this activity with another item such as number of books in each desk or number of buttons on clothing.

Reading Bar Graphs

Set up: Quickly sketch the bar graph below. Include the numbers along the axes.

1. **Several people in Scott's family carry keys: apartment keys, gym locker keys, mailbox keys, car keys. How many people in Scott's family carry keys?** (7 people)
   Tell me which axis to label "Number of People" and which to label "Number of Keys."

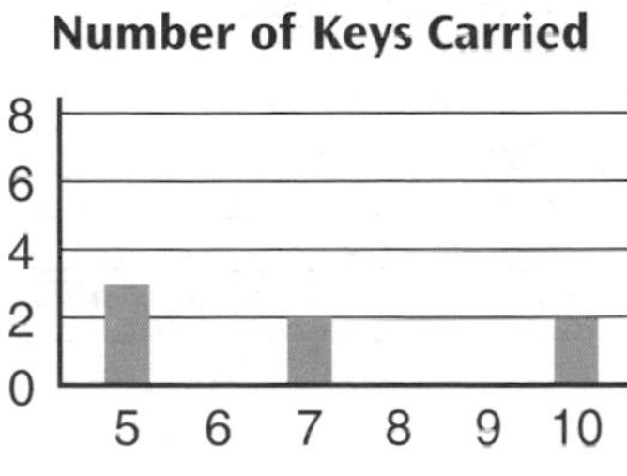

2. **How many people carry the fewest number of keys? How many do they carry?** (3 people, 5 keys)
   What is the greatest number of keys that anyone carries? (10 keys)
   What is the mode? (5 keys)

3. **Explain how to find the median. What is the median?** (7 keys)

4. **Explain how to find the mean. What is the mean?** (7 keys)

Data

Circle Graphs

Set up: Sketch two columns on the board. Students need slates or paper.

1. **Twelve players voted on uniforms. Four voted for sleeveless shirts, 6 voted for short sleeves and 2 voted for long sleeves.**
   Tell me how to create a data table. Be sure the table is titled and the columns are labeled.

2. Add two columns to the right of your table and label them "Fraction of Team" and "Percent of Team."
   Copy the table on your paper and complete it.

Shirts	Votes	Fraction of Team	Percent of Team
Sleeveless	(4)	($\frac{4}{12}$ or $\frac{1}{3}$)	($33\frac{1}{8}$%)
Short Sleeve	(6)	($\frac{6}{12}$ or $\frac{1}{2}$)	(50%)
Long Sleeve	(2)	($\frac{2}{12}$ or $\frac{1}{6}$)	($16\frac{2}{3}$%)

3. **Sketch a rough circle graph for the data.**
   Label each section and give the graph a title.

Data

Median and Mean

Set up: Students need calculators and slates or paper.

1. **Determine your age in months and write it down.**
   Let's take a sample of five ages.
   Ask five students to report their ages in months and record them on the board.

2. **Look at the five pieces of data we have.**
   How do we find the median of the data? What is the median?
   How do we find the mean? What is the mean?

3. **Let's add five more pieces of data.**
   Have five different students provide their ages.
   Now that we have new data, do you think we have a different median? Why?
   How do we find the median of the 10 ages? What is the median?
   Do you think we will have a different mean? Why? What is the mean of the data?

Stem-and-Leaf Plot

Set up: Write the numbers 130, 135, 126, 120, 131, 135, 138, and 128 on the board.

1. **On the board I've written the weights in pounds of female black bears at a zoo.**
   Make a stem-and-leaf plot of the weights of these bears.

2. **Find the median, minimum, range, and mode of the data.** (130.5, 120, 18, 135)

3. **Suppose that the zoo acquires two more female black bears. When the new bears' weights are added to the data set, none of the landmarks change.**
   Explain how that could happen.
   What might each new bear weigh? (Example: 129 and 132)

Source: Idaho Public Television, *www.idahoptv.org/dialogue4kids/bears/facts.html*

Data

Data with Negative Numbers

Set up: Draw the chart below on the board. Students need slates or paper.

J	F	M	A	M	J	J	A	S	O	N	D
–21	–15	–4	10	25	36	41	38	29	17	9	–1

1. **Record this data set on your paper or slate.**
   These are the lowest temperatures on record through 1998 in degrees Fahrenheit for each month in Detroit, Michigan. This data is based on 40 years of records.
   What are the minimum and maximum of this data set? (–21, 41)

2. **Rewrite the data, ordering it from lowest to highest temperature.**
   (–21, –21, –15, –4, –1, 9, 10, 17, 25, 29, 36, 38, 41)
   What is the range? (62) **How did you determine the range?**

3. **What is the median?** (13.5) **How did you determine the median?**

Source: Northeast Regional Climate Center, *met-www.cit.cornell.edu/ccd/lowtmp98.html.*

Data

Time Graph

Set up: Sketch two axes on the board. Students need slates or paper.

1. **We are going to sketch a graph that shows Brian's speed as he runs at the track.**
   I'm going to label the horizontal axis "Time."
   What should the vertical axis be labeled? (Example: "Speed")
   Draw and label the horizontal and vertical axes on your paper.

2. **Brian stands on the track waiting to start. Then, he takes off.**
   Where should the line start? (Example: At the origin)
   Show me with your hand whether the line stays flat, slants up, or slants down. (Slants up)
   Sketch a small portion of the curve.

3. **Brian runs halfway around the track, then walks halfway around, and then runs one complete lap around the track.**
   When he runs, he tries to run at the same speed.
   Continue the line on your graph to show Brian's speed as he exercises.

Data

Circle Graph

Set up: Write the following on the board: Oranges $\frac{1}{3}$, Bananas $\frac{1}{4}$, Pears $\frac{1}{6}$, and Apples. Students need slates or paper.

1. **At the class picnic there were 36 pieces of fruit.**
   How many oranges were there? Bananas? Pears? (12, 9, 6)
   How many apples were there? What fraction of the fruit was apples? (9, $\frac{1}{4}$)

2. **Sketch a rough circle graph that displays this data.**

Reusable: Repeat this activity with 12, 24, or 48 pieces of fruit.

Data

Reading a Time Graph

Set up: Sketch the distance-time graph shown below on the board. Students need slates or paper.

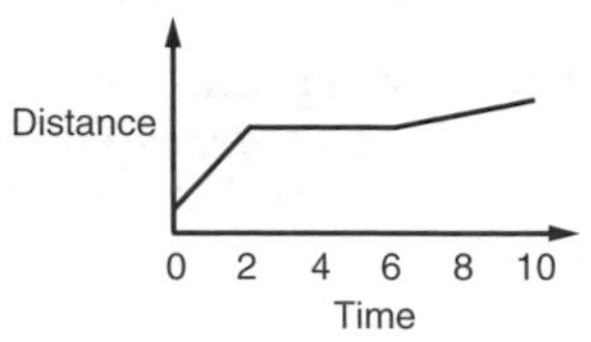

1. **Write a story that fits this graph.**

2. **Let's share some of our stories.**

Reusable: Repeat this activity with a different graph.

Data

Probability: Outcomes

Set up: Draw three squares on the board and have students draw three squares on their slates or a piece of paper. Give three students one die each.

1. Have one student roll a die and record the result in the first square.
   What was the chance of this number being rolled on the die? ($\frac{1}{6}$)
   Write the student's name above the square and $\frac{1}{6}$ below it. Continue with the second and third students rolling their dice, recording the results and their names, and indicating the chance.
   What do you notice about these chances?
   Discuss the idea of "equally likely" and why dice are used in many games.

2. **How can I determine the total number of possible outcomes when I roll 3 dice?**
   Have students provide ideas. Lead the discussion toward the Multiplication Counting Principle and the use of tree diagrams.

3. **If I roll these 3 dice again, what is the chance that I will get the same result as before?** ($\frac{1}{216}$)

Probability

Probability: Likelihood

Set up: The Probability Meter Poster is optional.

1. **When we talk about probability, we are talking about the chance that something will happen.**
   Give me some examples of a probability. (Example: A 50% chance of getting HEADS when flipping a coin)

2. **In these examples, there are levels of likelihood. What are some of the terms that we use to express the likelihood that something will happen?**
   (Examples: Certain, likely, 50-50 chance or equally likely, unlikely, impossible)

3. **Give me an example of something that is certain.** (Example: The sun will rise tomorrow.)
   What is the probability of this certain event? (1 or 100%)
   Give me an example of something that is impossible. (Example: I will roll a 7 with a six-sided die.)
   What is the probability of this impossible event? (0 or 0%)
   What are the minimum and maximum values for a probability? (0 and 1)

Vocabulary: Experiments, Outcomes, Events

Set up: Write Experiment, Outcomes, and Event on the board.

1. **Give me some examples of situations where we would talk about the probability of something.** (Example: Rolling a die)
   Choose one of the examples to explore.
   Tell me some different ways to express this probability. (Examples: fraction, decimal, percent)

2. Continue with one of the student examples or use the following:
   Look at the three terms on the board and listen to the following scenario: We roll a die and look at what number is on top. We want to know the probability of rolling an even number.
   What is the experiment? (Rolling a die)
   What are the possible outcomes? (1, 2, 3, 4, 5, 6)
   What is the event we are looking for? (An even number)
   How can our event occur? (Rolling 2 or 4 or 6)
   What is the probability of our event occurring? ($\frac{1}{2}$ or equivalent) **Why?**

Reusable: Repeat this activity to review probability terminology using a variety of experiments and associated events.

Doing an Experiment

Set up: Have three or more dice available. Draw a tally chart like the one below on the board. Students need slates or paper.

1. **What is the probability of getting a 2 when I roll this die?** ($\frac{1}{6}$ or equivalent)
   And the probability of getting a 5? Or a 3? Or a 1? (all $\frac{1}{6}$)
   So, we say that all of our outcomes are equally likely.

2. **If I roll this die six times, about how many 6s should I roll? 4s?** (1, 1)
   Let's see what happens. Within your group have one person roll the die 12 times and another person record the results.
   As groups finish, record the results on the board.

3. **Did any group roll exactly 2 of each number? Why not? Let's total all of the results. What do you notice?**

Group	1	2	3	4	5	6
1						
2						
3						
TOTAL						

Probability

Tree Diagram

Set up: Write the following on the board: Crust: thick or thin, Topping: pepperoni, mushroom, or sausage. Students need slates or paper.

1. **We want to order a one-topping pizza.**
   With a partner, begin to list all of the different types of one-topping pizzas we could order using the crusts and toppings shown on the board.
   Allow students a short period to start their lists and then stop them.
   How will you know when you have listed all of the possibilities?

2. **Let's use a tree diagram to show all the possible pizzas. Work with your partner to draw a tree diagram.**
   After students have worked for a minute, work as a whole class to draw the tree diagram on the board.

3. **How could we use our tree diagram to count the number of possible pizzas?**
   How could we use our tree diagram to list all of the possible pizzas?

4. **What would we have to do to our diagram if we wanted to add the option of "extra cheese" or "no extra cheese"?**

Multiplication Counting Principle

Set up: Write the following on the board: Cones: cake or sugar, Ice Cream: vanilla, chocolate, or strawberry. Students need slates or paper.

1. **We are going to have a party and serve single-scoop ice cream cones. How can we determine the number of different single-scoop ice cream cones we can make using the cone and ice cream choices shown?** Allow students to offer suggestions and record them on the board. Make sure that "carefully make a list," "use a tree diagram," and "use the Multiplication Counting Principle," or a description of these, are included in the list.

2. **Which of these methods is the quickest to determine how many different cones we can make? How many different cones can we make?** (6) **Which method will best help us list all of the possible cones?**

3. **Let's add a topping choice of either nuts or sprinkles to our cone. How do I determine the number of different cones I can make now?** (12) **Some people put their cone into a cup and eat it with a spoon. If we add "cup and spoon" or "no cup and spoon" to our choices, how many different single-scoop ice cream cone options do we have?** (24)

Conversion of Units

Set up: Students need slates or paper and meter sticks, yard sticks, or tape measures.

1. **On your slate or paper write an estimate of the length of your desk in millimeters.**
2. **Now verify your guess by measuring.**
   Discuss the variations that arise and why they may happen.
3. **Measure again but express the length in millimeters, in centimeters, and then as a part of a meter.**
   Write "Example: 610 mm = 61 cm = 0.61 m" on the board.

Reusable: Repeat this activity using the width of the desk or another object in the classroom.

Capacity

Set up: Students need slates or paper and calculators.

1. **The National Institutes of Health recommends that students from 9 to 18 years old drink four cups of milk a day. How many pints is that? How many quarts?** (2 pt, 1 qt)

2. **Now calculate the weekly recommendation in quarts for the entire class.** (Number of students *7)

3. **Cows produce an average of about eight gallons of milk per day. Decide whether one cow could provide enough milk for the class.** (Yes, 8 gallons = 32 quarts and students need 1 quart per day.)

Conversions: Yards, Feet, Inches

Set up: Draw a chart on the board with columns for yards, feet, and inches. Write the measures shown below in the yards column. Draw another chart on the board for use later. Students need slates or paper.

Yards	Feet	Inches
2	(6)	(72)
$1\frac{1}{3}$	(4)	(48)
$1\frac{1}{2}$	(4.5)	(54)
$1\frac{1}{6}$	(3.5)	(42)

Yards	Feet	Inches
(1)	(3)	36
$(\frac{2}{3})$	(2)	24
$(\frac{1}{3})$	(1)	12
$(\frac{1}{6})$	$(\frac{1}{2})$	6
$(\frac{1}{12})$	$(\frac{1}{4})$	3

1. **Find the foot and inch equivalence for each yard length shown in the chart. Show your work on paper or your slate.**
   Have selected students explain their thinking or show work at the board.

2. Write the following inch measures in the second chart: 36, 24, 12, 6, 3.
   Now fill in the missing foot and yard measures in this chart.

Measurement

Units of Weight: Conversions

Set up: Write "A typical 10-year-old weighs 77 pounds or 34.9 kg"[1] on the board. Calculators are optional. Students need slates or paper.

1. **Look at the weights on the board. How many ounces does a typical fifth grader weigh?** (1,232 oz)
   How many grams does a typical fifth grader weigh? (34,900 g)

2. **Roughly how many grams are in an ounce?** (28 g)
   How did you determine your answer? (Example: Fifth grader weight in grams divided by weight in ounces)

3. **An average baby elephant weighs 232 pounds at birth.**[2]
   How many ounces does an average baby elephant weigh? (3,714 oz)
   How many grams is that? (103,992 g)
   How many kilograms is that? (103.992 kg)

Sources: 1. Eric Schwarz, "Realizing the American Dream: Historical Scorecard, Current Challenges, Future Opportunities," working paper for *A Gathering of Leaders* leadership conference (Cambridge, Mass.: New Profit, 2005).
2. San Diego Zoo, *www.sandiegozoo.org/animalbytes/t-elephant.html*

Measurement

Circumference and Area

Set up: Draw the coins as shown. Calculators are optional. Students need slates or paper.

1. **Use the information on the board to estimate the circumference of each of our coins.** (using $\pi \approx 3$: penny, $\approx$ 57 mm; nickel, $\approx$ 60 mm; dime, $\approx$ 51 mm; quarter, $\approx$ 72 mm)

19 mm

20 mm

17 mm

24 mm

Diameters of Coins

2. **Use the information on the board to estimate the area of each coin.** (using $\pi \approx 3$: penny, $\approx 270\ mm^2$; nickel, $\approx 300\ mm^2$; dime, $\approx 217\ mm^2$; quarter, $\approx 432\ mm^2$)

3. **Which two coins have areas that are the closest in size?** (nickel and penny)

4. **Which coin has an area about four-fifths the area of another coin?** (The dime's area is about $\frac{4}{5}$ of the penny's.)

5. **Which coin's area is about 70% the area of another?** (The nickel's area is about 70% of the quarter's.)

Time Conversions

Set up: Draw the chart on the board. Students need slates or paper and calculators are optional.

	Minutes/Week	Hours/Month	Hours/Year
PE			
Math			

1. **Help me fill in the weekly estimates for PE and math on this chart. Think about how many times a week we have PE and math. About how many minutes of PE do you have in a week? About how many minutes of math do you have in a week?**
   Write a consensus answer on the board for each subject.

2. **Now work in pairs to fill in the remaining blocks on the chart. How did you determine your entries?**
   Discuss students' solutions.

Scale and Conversion

Set up: Write "1 inch stands for 1 foot" and then sketch a skateboard on the board. Students need slates or paper.

1. **This drawing of a skateboard is two and one-half inches long. The scale says that each inch of the drawing represents one foot of the real skateboard.**
   What is the length of the skateboard? (2.5 ft)
   How many inches is that? (30 in.)
   How many yards? ($\frac{5}{6}$ yd)

2. **Write down your approximate height in feet and inches and in inches. If you used the same scale and sketched a picture of yourself, how tall would you be in the picture?**

3. **Sketch a picture of yourself on a skateboard.**
   Try to show your height in comparison to the length of the skateboard.

Source: Skatesonhaight.com, *www.skatesonhaight.com/Birdhouse_skateboards_s/28.htm*

Measurement

Scale and Perimeter

Set up: Sketch the rectangle below and write "1 unit stands for 2 feet" on the board. Students need slates or paper.

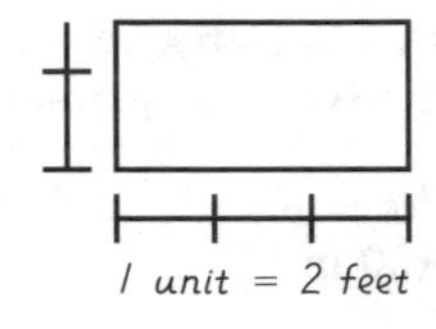

1. **Here is a scale drawing of a garden. Each unit represents two feet. What are the length and width of the garden in feet?** (6 ft; 3 ft)

2. **What is the perimeter of the garden in feet?** (18 ft)

3. **Make a scale drawing of a trampoline that is 13 feet by 14 feet.**

Measurement

Congruence

Set up: Two pieces of $8\frac{1}{2}$ by 11 inch paper and two rectangular books of different sizes are needed.

1. **What makes two figures congruent?**
   Make sure students conclude that figures are congruent when they are the same size and the same shape.

2. Hold up the two pieces of $8\frac{1}{2}$ by 11 inch paper.
   Are these two sheets of paper congruent? (Yes)
   Hold up the two rectangular, but different sized books.
   Are these two books congruent? (No)
   Repeat with other pairs of objects at hand, such as erasers, pencils, chalk, or desks.

3. **Look around the room and find some congruent shapes.**
   Share your ideas with the person next to you.
   What congruent objects did you and your partner find in our classroom? Explain your choices.

4. Remind students of the symbol for congruence ($\cong$).

Angles in Triangles and Quadrangles

Set up: Draw two triangles on the board, labeled as below. Students need slates or paper.

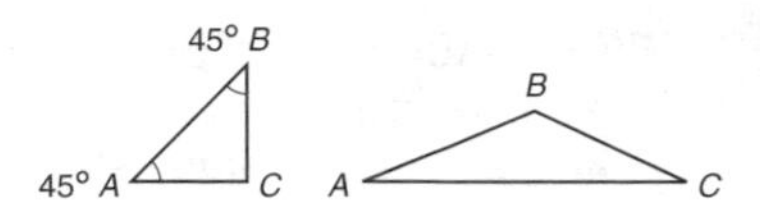

1. **The measures of angles *A* and *B* in this triangle are 45°.**
   What is the measure of angle *C*? (90°)

2. **Now let's look at another triangle. Give me some possible measures for angles *A* and *B*.**
   Accept two reasonable answers. Label the angles from student suggestions.
   What is the measure of angle *C*?
   In your groups, draw a new triangle *ABC*. Estimate measures for two of the angles and then find the third angle in the triangle.

3. Draw a quadrangle *ABCD* and label angles *B, C,* and *D* with different measures.
   Tell me the measure of angle *A*.
   Make up new measures for three angles. Find the fourth angle measure.

Reusable: Repeat this activity with different angle measures and different polygons.

Geometry

Regular Polygons

Set up: Students may need slates or paper.

1. **In some Kindergartens, when the teachers want to talk to all of the students, they ask them to sit in a circle on the floor.**
   If all the students in our class today were to sit on the floor at the same time, could we form an equilateral triangle with no one left out?
   How did you determine whether or not we could?
   Could we form a square?

2. **What regular polygons could our class form using all students?**
   Do you notice a pattern?

3. **What regular polygons could we form if I join you?**
   Continue by including absent students or the principal.

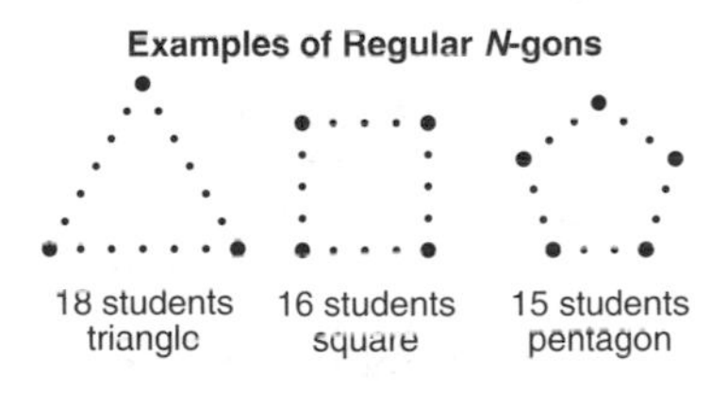

Geometry

Paper-Folding Angles

Set up: Students need an $8\frac{1}{2}$ by 11 inch sheet of paper and can work alone or in pairs.

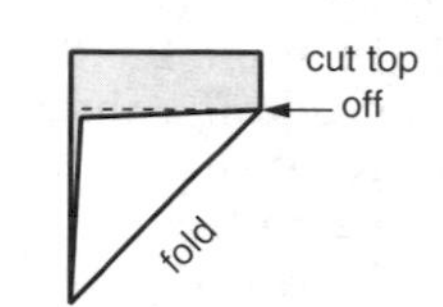

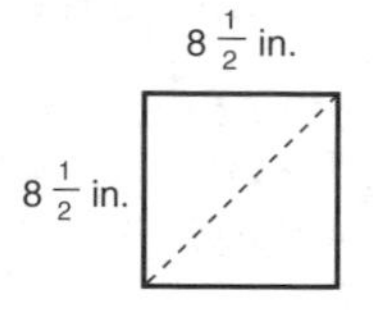

1. **First, we're going to make a square from this sheet of $8\frac{1}{2}$ by 11 inch paper. Line up the bottom edge of the paper with the lower part of the left edge like this. Help each other. Make a sharp crease.**
   Now fold the top of the paper over like this and make another sharp crease.
   Cut or tear off this top piece.
   Now you have a square. What is the fold in the square called? (diagonal)
   Find two different straight angles. How many degrees are in a straight angle?
   Write "1808 ∗ 2 = 3608" on the board.

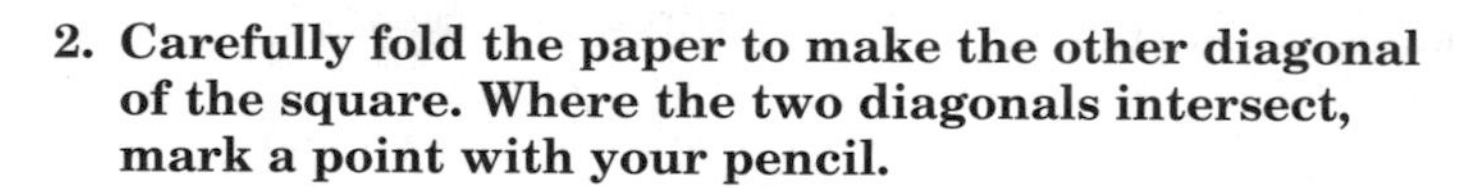

2. **Carefully fold the paper to make the other diagonal of the square. Where the two diagonals intersect, mark a point with your pencil.**

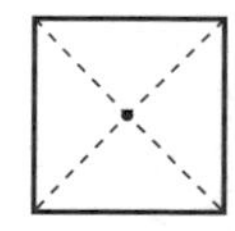

continued

3. **Your paper folding has created several angles around this center point. Can you identify the angles around the center point? Work with a partner to identify and label the four right angles.**
   As students share their results, write "90° * 4 = 360°" on the board.

4. **Now fold the square in half vertically and then horizontally.** (See illustration and demonstrate.)
   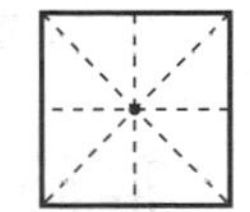
   Each of the right angles is bisected by a fold creating angles less than 90 degrees. How many of these are there and what do they measure?
   Write "45° * 8 = 360°" on the board.

Volume to Surface Area

Set up: Draw the table below on the board. Students work alone or with a partner and need slates or paper.

1. **Draw a table labeled "Cube" like the one on the board. We'll fill in the first row together. If a cube has a side length of 1, what is its volume? What is its surface area? So, the ratio of the volume to the surface area in this cube is what?**

Cube

Side Length	Volume	Surface Area	V : SA
1	(1)	(6)	(1 : 6)
2	(8)	(24)	(1 : 3)
3	(27)	(54)	(1 : 2)
4	(64)	(96)	(2 : 3)

2. **Now fill in the next three rows on the chart. Do you notice a pattern?**
   (As the size of the cube increases, so does the V : SA ratio.)
   Discuss student findings.

3. **Try some other numbers and check your predictions.**

Geometry

Classifying Quadrilaterals

Set up: Write the words *quadrilateral, trapezoid, parallelogram, rectangle, square,* and *rhombus* on the board. Students need a sheet of paper.

1. **Fold a piece of paper into six sections. Write the first word from the board in the first section, the second in the second section, and so on. In each section, draw two different representations of that shape.**
   Have students check each other's work before proceeding.

2. **In the trapezoid section, we can write, "Must have one pair of parallel sides." Now, in the parallelogram section, write a phrase that tells something a parallelogram must have that a trapezoid might not.**
   (Example: Two pairs of parallel sides)
   Have students check each other's work before proceeding.

3. **In the rectangle and square sections, write a phrase that describes something each shape must have that the previous shape might not.**
   (Example: Four right angles, four equal sides)

4. **Is a rhombus a trapezoid or a parallelogram? Why? Is every square a rhombus? Why or why not?**

Classifying Triangles

Set up: Draw the chart below on the board. Students need a sheet of paper.

Triangles by Sides	Triangles by Angles

1. **Triangles are named by their side length relationships and by the type of angles they have. Tell me some of the names we use for triangles.**
   (sides: scalene, isosceles, equilateral; angles: acute, right, obtuse)
   Collect terms from students and write them in the proper column.

2. **Fold a sheet of paper into six parts.**
   In each section, draw different types of triangles and name them.
   You can use one or two terms to describe each triangle; for example, scalene right triangle.

Properties of 3-Dimensional Shapes

Set up: Draw the Venn diagram below on the board. Students need slates or paper. Physical examples of each solid are optional.

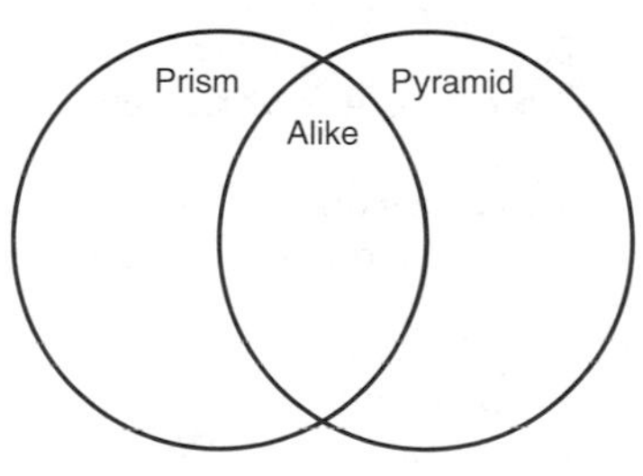

1. **On your paper, draw two large intersecting circles to make a Venn diagram like the one shown on the board. In the intersection of the circles, describe things that both a prism and a pyramid have in common.** (Example: They both have at least one base.)
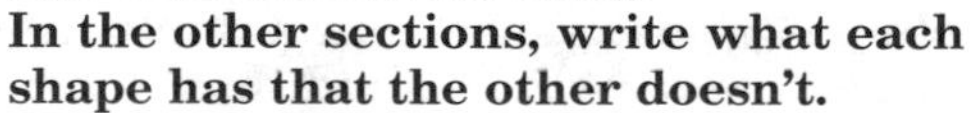
   In the other sections, write what each shape has that the other doesn't.
   (Examples: A prism has two bases. A pyramid has an apex.)
   You may want to have students offer an example for each section before students work independently.
   Collect students' ideas on the board before assigning #2.

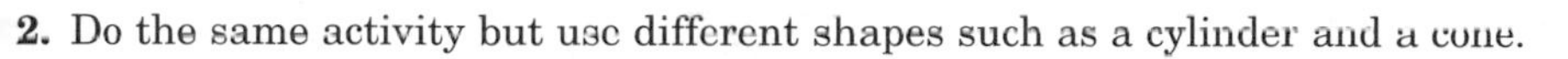

2. Do the same activity but use different shapes such as a cylinder and a cone.

Transformations

Set up: Students need paper. Graph paper, rulers, and transparent grid paper are optional. Write the following ordered pairs on the board: (0,3), (0,4), (1,4), (4,6), (4,1), (1,3), and (0,3).

1. **Fold a sheet of paper into fourths so the folds form the axes for a graph.**
   Darken the axes with your pencil.
   On your axes, plot the points shown on the board and connect them in order.

2. **Now, create a reflection of the shape by changing the first number in each pair to the opposite number, or negative value, and then graphing the new points.**

3. **On the back of the paper, create your own reflection picture.**
   Keep it simple—no more than 10 points on one side of the axis.

Symmetry: Line of Reflection Outside the Figure

Set up: Draw $\triangle ABC$, $\overleftrightarrow{MN}$, and $\overleftrightarrow{OP}$ on the board as shown below. Students need slates or paper.

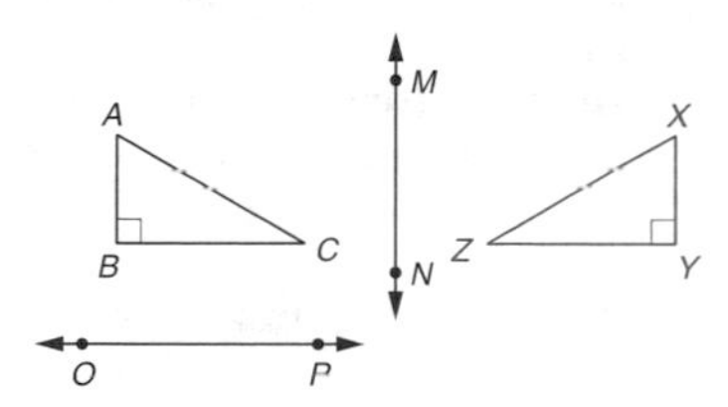

1. **Reflect $\triangle ABC$ in $\overleftrightarrow{MN}$.**
   Draw the reflected image.
   Describe the placement of the hypotenuse.

2. **Label the reflected image as $\triangle XYZ$.**
   What point in the reflected image is the same distance from $\overleftrightarrow{MN}$ as point A? (X)

3. **Now reflect $\triangle ABC$ in $\overleftrightarrow{OP}$.**
   Draw the reflected image and label it as $\triangle SRT$.
   Describe the preimage and the image.

Geometry

Writing Algebraic Expressions

Set up: Students need slates or paper. Write $x + 2$ on the board.

1. **Here is a story that fits with $x + 2$: Shawn is 2 years older than his brother.**
   In this story, what does $x + 2$ represent? (Shawn's age)
   Work with a partner and think of two additional stories that also fit the expression $x + 2$.

2. **Write an algebraic expression for each statement.**
   A teacher is x years older than her 11-year-old students. $(11 + x)$
   Gym class is held twice a week during y weeks of the school year, except for 5 days of assemblies. $(2y - 5)$

3. **Write an expression for how many pencils are given out if each student in your class receives m pencils.**
   Calculate how many pencils will be given out if $m = 2$.

Equivalent Expressions

Set up: Write the numbers 2, 4, 6, and 8 and the operation symbols +, −, *, and / on the board. Students need slates or paper.

1. **On your slates or paper, using the numbers 2, 4, 6, 8, and the operations, write at least two equations that equal 10.** (Examples: $6 - 4 = 10$; $8 + 2 = 10$)
   Compare your equations with each other.

2. Add the parentheses symbols ( ) to the board.
   Using the numbers 2, 4, 6, 8, the operations, and parentheses, write at least two equations that equal 100. (Examples: $(2 - 8)(6 - 4) = 100$; $(6 * 8 * 2) + 4 = 100$)
   Compare your equations with each other.

3. **Using the numbers 2, 4, 6, 8, the operations, parentheses, and any exponent, write an equation that equals 1,000.**
   (Examples: $(2 - 8)^2 * (6 - 4) = 1{,}000$; $(6 - 4)^3 = 1{,}000$)
   Compare your equations with each other.

Algebra

Problems with Variables

Set up: Draw a two-column table on the board. Label the first column "Yards" and the second column "Feet." Students need slates or paper.

1. **How many feet are in 6 yards? 7 yards? 8 yards?** (18 ft; 21 ft; 24 ft)
   Record this information in your table.
   What is the general pattern? (The number of feet is three times the number of yards.)
   How many feet are in 13 yards? (39 ft)
   Write a rule for finding the number of feet if you know the number of yards. Let f stand for feet and y stand for yards. ($f = 3y$)
   Use your rule to find the number of feet in 71 yards. (213 ft)

2. **Write a rule for the reverse, finding the number of yards if you know the number of feet.** ($y = f / 3$)

3. **There are 32 tablespoons in 2 cups. Write a rule with the variables c and T for finding the number of cups when you know the number of tablespoons.** ($c = T / 16$) **Write a rule for the reverse, finding the number of tablespoons when you know the number of cups.** ($T = 16c$)

Patterns

Set up: Draw the *in/out* table on the board.

1. **Work with a partner. Copy the table and complete it.**

2. **Use words to write a rule for the *in/out* table.** (Example: The *out* number is 1 less than 2 times the *in* number.)
   Suppose a mystery number M is in the *in* column. How can we determine the output? (Example: Multiply M by 2 and then subtract 1.)
   If there is a mystery output Q, how can we determine what the input was? (Example: Add 1 to Q and divide the total by 2.)

in	*out*
6	11
3	5
(2)	3
10	(19)
(50)	99
0	(−1)
8.5	(16)
(3.5)	6

Number Patterns

Set up: Students need slates or paper. Write the numbers 4, 5, 6, 7, and 8 on the board.

1. **How many pairs of whole numbers will sum to 6?** (4 pairs: 0 + 6, 1 + 5, 2 + 4, 3 + 3)
   Note that 6 + 0 and 0 + 6 are the same pair of numbers.
   Write the pairs on the board.

2. **Work with a partner and determine how many pairs of numbers sum to each of the numbers on the board.** (For 4: 0 + 4, 1 + 3, 2 + 2; For 5: 0 + 5, 1 + 4, 2 + 3; For 6: 0 + 6, 1 + 5, 2 + 4, 3 + 3; For 7: 0 + 7, 1 + 6, 2 + 5, 3 + 4; For 8: 0 + 8, 1 + 7, 2 + 6, 3 + 5, 4 + 4)

3. **What patterns do you notice?** (Examples: As the sums increase by 2, the number of pairs increase by 1; the number of pairs for an even number, E, is half E plus 1; the number of pairs for an odd number, Q, is the sum $Q + 1$ divided by 2.)
   How many pairs of whole numbers will sum to 9? To 10? (5, 6)

Algebra

Whole Number Sums and Differences

Set up: Sketch a two-column table on the board. Students need slates or paper.

1. **Tell me a pair of whole numbers whose sum is 12.** (Example: 4 and 8)
   What other pairs of whole numbers have a sum of 12?
   Fill in the chart as pairs are suggested.
   How many pairs of whole numbers have a sum of 12? Explain. (7)

2. Draw a new two-column chart next to the first chart.
   Tell me a pair of whole numbers whose difference is 12.
   (Example: 13 and 1)
   Write them in the chart.
   Work with a partner and find several pairs of whole numbers whose difference is 12.
   How many pairs of whole numbers do you think have a difference of 12? Explain. (Infinite. Every number is 12 more than some other number.)

3. **What patterns do you notice in these charts?**

Reusable: Repeat this activity using numbers less than 12 for the sum and difference.

Formula

Set up: Write $P = 4 * S$ on the board.

1. **This formula can be used to find the perimeter of a square. We let *P* be the perimeter and *S* be the length of one of the sides of the square. What is the perimeter of the square if the length of one side is 20 inches? 9 feet? 7 yards?** (80 in., 36 ft, 28 yd)

2. **How can you find the length of the sides of a square if you know that the perimeter is 12 feet? 9 inches? 28 yards?** (Divide the perimeter by 4: 3 ft, $2\frac{1}{4}$ in., 7 yd)
   Write a formula to find the side *S* if you know the perimeter *P*. ($S = P / 4$)

3. Sketch the two-column table at right on the board. **Tell me the missing values in this table.**

Perimeter of Square (ft)	Side of Square (ft)
20	(5)
10	($2\frac{1}{2}$)
(16)	4
5	($1\frac{1}{4}$)

Sequence: Square Numbers

Set up: Write 1, 4, 9, 16,...on the board. Students need slates or paper.

1. **Jeff thought of a rule and then wrote this sequence of numbers. What do you think his rule was? Explain.** (They are square numbers: 1^2, 2^2, 3^2, 4^2.)

2. **What are the next three numbers in this sequence? Explain.** (25, 36, 49. They are 5^2, 6^2, and 7^2.)

3. **Work with a partner and find the tenth number in the sequence.** (100)

Algebra

Sequence

Set up: Write __, 5, 10, 17, __, 37, __,...on the board. Students need slates or paper.

1. **Sue thought of a rule and then wrote this sequence of numbers. What do you think her rule was? Explain.** $(x^2 + 1)$
   Can you find another rule that would also work? $((x * x) + 1)$

2. **What are the missing numbers? Explain.** (2, 26, 50)

3. **Work with a partner and determine if 100 is a number in the sequence. How did you decide?** (No, because $100 - 1$ is not a square number.)

Reusable: Repeat this activity with any sequence and missing numbers in the sequence.

Function Machine: $\frac{n}{2} + 1$

Set up: Write $\frac{n}{2} + 1$ on the board. Draw an *in/out* table on the board as shown below. Students need slates or paper.

1. **Here is the rule for a function machine. The natural numbers 1, 2, 3, . . . are put into the machine in order. What is the first number that will come out of the machine?** (1.5)
   Fill in the table. Write 2, 3, 4, 5, and 6 in the *in* column. Write the next five numbers that will come out of the machine.

in	*out*
(1)	(1.5)
(2)	(2)
(3)	(2.5)
(4)	(3)
(5)	(3.5)

2. **What will be the tenth number to come out?** (6)

3. **Predict whether 15 can be an output number. Explain.** (Yes. Every whole number greater than 1 can be an output number.)

Functions

Set up: Draw the chart below on the board. Students need slates or paper.

1. **Copy and complete the chart shown on the board.**
2. **Write a function using x and y that shows the rule you used to fill in the chart.** (Examples: $y = 2x - 2$; $y = 2(x - 1)$)
3. **Fill in four more rows of the chart.**

x	y
1	0
(2)	2
3	4
(4)	6
(5)	8
6	10

Number Line

Set up: Draw a blank number line on the board. List the numbers 7, −3, 0, −1, 2, −6, and 4 on the board. Students need slates or paper.

1. **Draw a blank number line like this one on your paper.**
   Write these numbers in the appropriate places on your number line.
   Fill in the rest of the number line on the board using student suggestions.

2. **Let's use our number line to do some addition and subtraction problems.**
   Use the number line to show the following addition and subtraction problems. You may want to use two colors of chalk to indicate addition (moving along the line in a positive direction) and subtraction (moving in a negative direction).

   $1 - 3$ $\quad$ $-2 + 4$ $\quad$ $0 - 5$

Patterns

Set up: Draw a two-column table on the board. Label the columns x and y.

x	y
1	2
2	4
3	6

1. **Make a table that shows the numbers associated with this pattern: x is 1, 2, and 3 and y is 2, 4, and 6. If I put a 4 into the table for x, what is y?** (8)
   Go over the number pairs and extend the table.

2. **Write a rule in words that would create this table.**
   (Example: y is twice as much as x.)
   Now, write the rule as a formula. (Examples: $y = 2x$; $y = x + x$)
   Discuss student responses.

Reusable: Repeat this activity with other rules and formulas.

Algebra

Grouping Symbols

Set up: Write the problems in #1 on the board. Students need slates or paper.

1. **Solve the problems on the board and then we will discuss them.**
   $((8 + 3) \times 5) - 1 = (54)$
   $(8 + (3 \times 5)) - 1 = (22)$
   $8 + ((3 \times 5) - 1) = (22)$
   $8 + (3 \times (5 - 1)) = (20)$

   What do you notice about these problems? (The order in which the operations are carried out determines the result.)

2. **Make up your own problem like this and trade with a partner to solve. Check each other's answers.**

Associative Property

Set up: Write the problems under #1 and #2 on the board. Students need slates or paper.

1. **Copy the equations in this first group of problems and solve them.**

   $4 + (2 - 3) = (9)$ $(4 + 2) + 3 = (9)$ $4 + 2 + 3 = (9)$

   What do you notice? (The numbers can be added in any order. The sum is the same.)

2. **Now solve the equations in the remaining groups of problems.**

   $6 - (1 - 1) = (6)$ $10 \div (5 \div 5) = (10)$ $7 \times (2 \times 2) = (28)$

   $(6 - 1) - 1 = (4)$ $(10 \div 5) \div 5 = (\frac{2}{5})$ $(7 \times 2) \times 2 = (28)$

   $6 - 1 - 1 = (4)$ $10 \div 5 \div 5 = (\frac{2}{5})$ $7 \times 2 \times 2 = (28)$

   What appears to be true for addition and multiplication of several numbers that doesn't seem true for subtraction and division? (The order of operation only matters with subtraction and division.)

3. **Test some more equations like these and see if you get the same results.**

Algebra

Fraction Equivalents, Adding Fractions

Set up: Write $\frac{3}{8}$ on the board.

1. **Tell me a fraction that is equivalent to $\frac{3}{8}$.**
   Write the fraction on the board.
   Another one? (Examples: $\frac{6}{16}$, $\frac{12}{32}$)
   Repeat until you have written about 10 fractions equivalent to $\frac{3}{8}$.

2. **We want to add a fraction with the same denominator to each fraction on the board. The sum of both fractions should equal 1. What are the other fractions?** (Examples: $\frac{6}{16} + \frac{10}{16}$, $\frac{12}{32} + \frac{20}{32}$)
   For example, $\frac{3}{8} + \frac{5}{8}$ equals 1.

3. **If you add up four of the fractions from Problem 2, what is the smallest sum you can get?** ($1\frac{1}{2}$)
   What is the largest sum you can get? ($2\frac{1}{2}$)
   Tell me four fractions whose sum is 2. (All sums of four fractions should equal 2.)

Reusable: Repeat this activity using a different starting fraction such as $\frac{2}{5}$, $\frac{4}{7}$, or $\frac{5}{9}$.

Powers of Ten: Numbers Less Than or Equal to 1

Set up: Start a three-column chart and write 10^{-2} in the third row, first column (leaving the top two rows and headings empty).

(Power of 10)	(Decimal)	(Fraction)
(10^0)	(1)	$(\frac{1}{1}, \frac{1}{10^0})$
(10^{-1})	(0.1)	$(\frac{1}{10}, \frac{1}{10^1})$
10^{-2}	(0.01)	$(\frac{1}{100}, \frac{1}{10^2}, \frac{1}{(10*10)})$
(10^{-3})	(0.001)	$(\frac{1}{1{,}000}, \frac{1}{10^3}, \frac{1}{(10*10*10)})$
(10^{-4})	(0.0001)	$(\frac{1}{10{,}000}, \frac{1}{10^4}, \frac{1}{(10*10*10*10)})$

1. **How can I write this number as a decimal?** (0.01)
   How can I write it as a fraction? (Example: $\frac{1}{100}$) **Is there another way?** Write responses in the second and third columns.
   Tell me how I could label these columns. ("Power of Ten," "Decimal," "Fraction")

continued

2. **Don't call out, but show me by holding up fingers: What negative exponent is used to write one one-thousandth as a power of 10?** (3) Survey student answers and write 10^{-3} in the fourth row of the chart.
   How is this notation read? ("ten to the negative third")
   How should we write it in decimal notation? (0.001)
   In fraction notation? (Example: $\frac{1}{1,000}$)
   Complete the fourth row of the chart with students.

3. **Now hold up fingers to show which negative exponent is used to write one ten-thousandth.** (4) Write 10^{-4} in the fifth row of the chart.
   Tell me how to complete this row of the chart.

4. **What patterns do you notice in the chart?**

5. **Using the patterns you found, complete the chart, including the top row.**

Roman Numerals: I–CXV

Set up: Draw two columns on the board. Label the left column Roman Numerals and the right column Standard Notation. Fill in as illustrated below.

1. **Roman Numerals are an old system of writing numbers. You sometimes see them on monuments or buildings. What do think X equals?** (10) **What does XL equal?** (40) **What about LX?** (60) **XXX?** (30) **Look at 10, 20, and 30. Do you think that 50, 60, 70, and 80 follow a similar pattern? Tell me how to write them.** (L, LX, LXX, LXXX)

Roman Numerals	Standard Notation
I	1
X	(10)
XX	20
XXX	(30)
XL	(40)
L	50
LX	(60)

2. **C equals one hundred. What does XC equal?** (90) **What about CX?** (110)

3. Write V, XV, XXV, . . . CXV on the board.
We know that V equals five. Read this sequence of numbers and tell me how to continue it to CXV using Roman Numerals. (XXXV, XLV, LV, LXV, LXXV, LXXXV, XCV, CV, CXV)

4. Write LIV and LVI on the board.
How are these numbers written in standard notation? (54, 56)

Reusable: Repeat this activity using Roman Numerals to write sequences of numbers from 9 to 119 or from 7 to 127.

Roman Numerals: I–CXV (Review)

Set up: Students need slates or paper.

1. **What symbols are used to write Roman Numerals for numbers from 1 to 100?** (I, V, X, L, C)
   What is the value of each of these symbols? (1, 5, 10, 50, 100)
   What is the least (smallest) number that can be written in Roman Numerals with an X? (IX, 9)
   What is the greatest (largest) number between 1 and 100 that can be written in Roman Numerals with an X? (XCVIII, 98)
   What is the least number that can be written in Roman Numerals with an L? (XL, 40)
   What is the greatest number between 1 and 100 that can be written in Roman Numerals with an L? (LXXXIX, 89)

2. **Write three different, three-symbol Roman Numerals using the symbols I, X, and L just once in each number.** (XLI, LIX, LXI)

3. **What is the largest number you can write with a three-symbol Roman Numeral that uses each of the symbols I, X, and C just once?** (CXI, 111)
   What is the smallest number? (XCI, 91)

Scientific Notation: Very Small Numbers

Set up: Write 0.0006 m on the board.

1. **Do you know what a hyphen is? This measure is a bit less than the length of a printed hyphen.**
   Read this measurement. What does the *m* stand for? (meter)
   How do I write "six ten-thousandths of a meter" in scientific notation?
   How would I read this? ($6 * 10^{-4}$ m, "six times ten to the negative four meters")

2. **Certain bacteria that live in surgeonfish are about this size. This is unusually large for bacteria.**
   Some bacteria are much smaller, about one ten-millionth of a meter.
   How do I write one ten-millionth of a meter in scientific notation?
   How do I write it in standard notation? ($1 * 10^{-7}$ m, 0.0000001 m)

3. Point to the $6 * 10^{-4}$ m and the $1 * 10^{-7}$ m on the board.
   Which is the longer measurement? ($6 * 10^{-4}$ m)
   Now tell me a measurement between the two. Be sure to tell me how to write the measurement in scientific notation. (Example: $2 * 10^{-5}$ m)

Source: Bacteriophage Ecology Group, *www.mansfield.ohio-state.edu/~sabedon/biol2010.htm*

Divisibility Rules for 3 and 9

Set up: None

1. **When we divide a number by 3 and the remainder is 0, then the number is said to be divisible by 3.**
   Is the number of students in the class divisible by 3? Is the number of eyes in the class plus the number of noses divisible by 3? (yes) **Can anyone tell me a shortcut for knowing when a number is divisible by 3?**
   If no one describes this shortcut tell it to them:
   Add up the digits in the number. If the sum is divisible by 3, then the number itself is divisible by 3.
   Let's try a few. How about 81? (yes) **How about 123?** (yes) **And 1,496?** (no)

2. **What do you think is a shortcut for knowing when a number is divisible by 9?**
   Write suggestions on the board. Add the following shortcut if no one suggests it:
   Add up the digits in the number. If the sum is divisible by 9, then the number itself is divisible by 9.
   Let's test this out. How about 81? (yes) **And 123?** (no) **And 1,494?** (yes)

Ordering Fractions

Set up: The board is needed. Students need slates or paper.

1. Ask 6 students to stand up in front of the class.
   What fraction of the class is standing up?
   What is the numerator? (6) **What is the denominator?**
   Write the fraction on the board. Ask one student to sit down.

2. **What fraction of the class is now standing? How do I write this?**
   If 2 sitting students leave, what fraction of the class will be standing?
   Write both fractions on the board. Ask all students to sit down.

3. **Order the three fractions on the board from largest to smallest.**

4. **A word is missing from the following statements. If you think the missing word is "increases," show me a thumbs-up. If you think the missing word is "decreases," show me a thumbs-down.**
   If the numerator increases but the denominator stays the same, then the fraction _______. (increases)
   If the numerator stays the same but the denominator increases, then the fraction _______. (decreases)

Fractions: $<1, = 1, >1$

Set up: Write the numbers 2, 4, and 7 on the board. Students need slates or paper.

1. **Using only these numbers, write as many different proper fractions as you can.** ($\frac{2}{4}, \frac{2}{7}, \frac{4}{7}$)
   After a minute ask students to tell you the fractions and write them on the board. **Write the fractions in simplest form.** ($\frac{1}{2}, \frac{2}{7}, \frac{4}{7}$) **Tell me how to order them from smallest to largest.** ($\frac{2}{7}, \frac{1}{2}, \frac{4}{7}$) **Find a common denominator and check that the fractions are in the correct order.** (Example: $\frac{4}{14}, \frac{7}{14}, \frac{8}{14}$)

2. **Using 2, 4, and 7, write all the improper fractions you can.**
   Be sure students include $\frac{2}{2}, \frac{4}{4}$, and $\frac{7}{7}$. ($\frac{2}{2}, \frac{4}{4}, \frac{7}{7}, \frac{7}{2}, \frac{7}{4}, \frac{4}{2}$)
   Write the fractions in simplest form. ($1, 1, 1, 3\frac{1}{2}, 1\frac{3}{4}, 2$) **Which fractions can be renamed as mixed numbers?** ($\frac{7}{2}, \frac{7}{4}$) **Tell me how to order all the proper and improper fractions from smallest to largest.** ($\frac{2}{7}, \frac{1}{2}, \frac{4}{7}, \frac{2}{2}, \frac{4}{4}, \frac{7}{7}, \frac{7}{4}, \frac{4}{2}, \frac{7}{2}$)
   Write them on the board.

Reusable: Repeat this activity with other sets of numbers, such as 3, 5, and 11; or 2, 3, and 9.

Negative and Positive Integers

Set up: Write "Location relative to sea level" on the board.

1. **Name situations where negative numbers might be used.**
   I have written an example on the board.
   Record students' ideas on the board.

2. **Use each of your ideas and write a description of a situation that could be represented with a –5.**
   In the sea level example, we could say "5 feet below sea level."
   Have students share their descriptions.

3. **Give a description of the opposite situation for each of your examples.**
   In our sea level example, we could say "5 feet above sea level."
   Have students share their descriptions.

Integer and Fraction Properties

Set up: Write the following list of properties on the board.

$$A + B = B + A$$
$$A - (-B) = B - A$$
$$A + (B + C) = (A + B) + C$$
$$A + (B - C) = (A + B) - C$$
$$A(B - C) = A(B) - C$$

1. **Some of the equations shown on the board work for all whole numbers.**
   Find the ones that do not work by citing a counterexample. In other words, replace the letters with numbers to show the equation is false.
   (Example: $2 - (-1) \neq 1 - 2$)

2. **Find the equations that do not work for fractions by citing a counterexample for each one.**

3. **Find the equations that do not work for integers by citing a counterexample for each one.**

Whole Number Theory

Set up: Write “Primes between 1 and 26” and “Composites between 1 and 26” on the board. Students need slates or paper.

1. **Prime numbers are whole numbers that have only two factors, one and the number itself.**
   List all the prime numbers between 1 and 26. (2, 3, 5, 7, 11, 13, 17, 19, 23)

2. **Prove that the remaining numbers are composite by showing that each composite number has a factor pair other than one and the number itself.** (Example: $4 = 2 * 2$)

Numeration

Decimal Rounding and Comparing

Set up: Draw the chart below on the board. Students need slates or paper.

Nearest Tenth Is 0.1	Nearest Hundredth Is 0.01	Nearest Thousandth Is 0.001
0.11	(0.008)	(0.0005)
(0.12)	(0.009)	(0.0006)
(0.101)	(0.011)	(0.0007)
(0.091)	(0.0105)	(0.0012)

(Answers shown in chart are examples only.)

1. **Copy this chart. In each box, list at least five numbers that could be rounded to the number shown. I have done one example.**
   Allow students to work and then write some student solutions in each box.

2. **Now let's put the numbers from each box in order from smallest to largest.**

Percent: Mental Calculation

Set up: Draw the first chart on the board. While students are working on the first chart, you can draw the second chart. Students need slates or paper.

1. Copy this chart and fill in the results. Try to do them in your head.

Beginning Amount	Percent	Result
$360	25%	($90)
$360	75%	($270)
$360	10%	($36)
$360	15%	($54)

2. Now find the sale price of a $360 television given the discounts shown. Again, try to do these in your head.

Beginning Amount	Discount	Sale Price
$360	5% off	($342)
$360	20% off	($288)
$360	15% off	($306)
$360	30% off	($252)

3. Describe the techniques you used to do these problems in your head.

Ordering Fractions

Set up: Draw the chart below on the board. Students need slates or paper. Calculators are recommended for #3.

$\frac{1}{2}$	$\frac{1}{3}$	$\frac{3}{4}$	$\frac{12}{5}$
$\frac{7}{12}$	$(\frac{10}{27})$	$(\frac{20}{25})$	$(\frac{16}{7})$

1. **Copy this chart. List five fractions that are close, but not equal to, the fraction heading each column. I have done one example.**

2. **Now put the fractions you named in each column in order from smallest to largest.**

3. **Check your ordering by converting the fractions to decimals.**

Fractions: Converting to Decimals

Set up: Write $\frac{1}{2}, \frac{1}{3}, \frac{1}{4}, \ldots, \frac{1}{10}$ on the board. Students need calculators.

1. **Who can remind us what it means for a fraction to have a terminating decimal equivalent?** (Example: The decimal equivalent is not repeating.)
   Which of the unit fractions shown on the board will produce a terminating decimal when entered into a calculator? ($\frac{1}{2}, \frac{1}{4}, \frac{1}{5}, \frac{1}{8}, \frac{1}{10}$)
   Which will not? ($\frac{1}{3}, \frac{1}{6}, \frac{1}{9}$)
   Are there clues that can help predict which fractions will terminate?

2. **Try some other unit fractions to find or test a theory.**
   Have students share ideas.

Estimation, Multiplication, Percent

Operations

Set up: Students may need slates or paper. Sketch the drawings below on board.

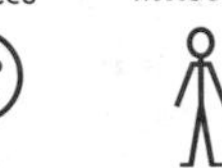

1. **Did you know that we use 14 muscles when we smile? If the entire class is smiling, how many muscles are being used? How did you determine your answer?** (Example: 14 * number of students)

2. **We each use 43 muscles to frown. Estimate how many muscles are being used if the entire class is frowning.**
How did you determine your answer? (Example: number of students * 40)

3. **There are about 650 muscles in the human body. If 1% of the muscles are at rest, about how many muscles is that?** (about 7)
How did you estimate to get your answer? (Example: 1% of 650 is 6.5. I rounded up to 7.)

4. **Remember that we use 14 muscles to smile. About what percent of our 650 total muscles do we use to smile?** (about 2%)

Division Word Problems, Decimals

Set up: The board is needed. Students need slates or paper.

1. **Make up a division problem with a 2-digit divisor and a quotient less than one.** (Example: 5 / 10)
   List some responses on the board.

2. **Make up two different word problems that go with your division problem.**
   Have students share problems.

3. **Change your division problem so the quotient is less than one-tenth.**
   (Example: 5 / 51)

Reusable: Repeat this activity with different requirements, such as finding quotients less than one hundredth or close to two tenths.

Fraction Addition and Subtraction Word Problems

Set up: Write $2\frac{1}{4} + 2\frac{1}{4} = 4\frac{1}{2}$ on the board. Students need slates or paper.

1. **Make up three more addition problems that have a sum of $4\frac{1}{2}$. One example is on the board.** (Examples: $4 + \frac{1}{2}$, $3 + 1\frac{1}{2}$, $3\frac{1}{2} + 1$)
   Write a word problem for one of your addition problems. (Example: Bob has $3\frac{1}{2}$ jars of marbles. If he gets another jar, how many will he have in total?)

2 **Make up three subtraction problems that have a difference of $4\frac{1}{2}$.** (Examples: $5\frac{1}{2} - 1$, $6 - 1\frac{1}{2}$, $5 - \frac{1}{2}$)
   Write a word problem for one of them. (Example: April is selling cups of lemonade. If she pours five cups, and then drinks half of one, how many cups does she have left?)

Fraction Multiplication and Division

Set up: Write the following on the board: ring $5\frac{2}{3}$ in., bracelet $12\frac{1}{4}$ in., pair of earrings $6\frac{3}{8}$ in. Students need slates or paper.

1. **A jeweler sells handmade rings, bracelets, and earrings. I've written on the board how much gold wire he needs to make each item. If gold wire comes in 3-foot lengths, how many rings could be made from each length of wire?** (6) **How many bracelets?** (2) **How many pairs of earrings?** (5)

2. **How many lengths of wire would be needed for four of each item?** (You would need 1 length of wire for 4 rings, 2 for 4 bracelets, and 1 for 4 pairs of earrings.)

3. **If you were the jeweler, which item would you focus on making and why?**

Operations

Operations

Decimal Multiplication

Set up: Write 4 * 0.25 = 1 on the board. Students need slates or paper.

1. **Make up four decimal multiplication problems where the product is exactly 1. One example is on the board.** (Examples: 0.25 * 4, 0.1 * 10, 0.5 * 2, 0.2 * 5)
   Write a word problem for one of your examples.

2. **Make up three decimal multiplication problems where the product is exactly 0.5.** (Examples: 0.05 * 10, 1 * 0.5, 0.25 * 2)
   Write a word problem for one of your examples.

Averaging Percents

Set up: Write the numbers 10, 15, 20, 25, and 28 on the board.

1. **Let's say we had a test with 10 problems and you earned an 80%. How would you determine the number of correct responses you had?** (Example: 0.8 * 10 = 8)

2. **If the numbers on the board are the number of items on your tests this marking period, how many items will you need to do correctly on each test to score at least 80% on each test?** (8, 12, 16, 20, 23)
   Discuss methods for finding at least 80% for each test.

3. **How many correct items would you need to earn a 90% on each test?** (9, 14, 18, 23, 26)

4. **Is there a way you could still have a 90% average if you scored only 80% on two of the tests?** (Example: Yes. You could also score 100% on two tests.)

Operations

Fraction Division

Set up: Give the problems below orally or write them on the board one at a time by changing only the numbers in the sentence.

1. **How many times can you subtract $\frac{1}{2}$ from 4?** (8)
   How many times can you subtract $\frac{1}{2}$ from 6? (12)
   How many times can you subtract $\frac{2}{3}$ from 6? (9)
   How many times can you subtract $\frac{2}{3}$ from 18? (27)
   Is there a quick way to do these subtractions? (Example: Think what number multiplied times the fraction gives the whole number.)

2. **Try your theory with mixed numbers.**
   How many times can you subtract $\frac{2}{3}$ from $5\frac{1}{3}$? (8)
   How many times can you subtract $\frac{3}{4}$ from $5\frac{1}{4}$? (7)

3. **Can you find another method that seems to work in all cases?** (Example: Invert the fraction and multiply.)

Properties of Integers

Set up: Write the following properties on the board:

$A + 0 = A$ $\quad$ $A + B = B + A$ $\quad$ $A + (B + C) = (A + B) + C$

1. **These properties are true for positive integers and whole numbers. Can you tell me in your own words what the first one is saying? And the second? The third?**

2. **Do you think these properties work for the negative integers? Use –4, –5, and –6 to test each one.** (yes)

3. **Do you think the properties will work with a combination of negative and positive integers? Use 4, –5, and –6 to test each one.** (yes)
   You can then review names of the properties of addition: Additive Identity, Commutative Property of Addition, Associative Property of Addition.

Reusable: Repeat this activity for the unit fractions and other fractions. Use $\frac{1}{2}$, $\frac{1}{3}$, and $\frac{1}{4}$ to test the properties for unit fractions. Use $\frac{2}{3}$, $\frac{3}{4}$, and $\frac{5}{6}$ to test them for other fractions.

Properties of Addition (and Non-Properties)

Set up: Students need slates or paper. Write the following equations on the board:

$A + 0 = A$	$A - B = B - A$
$A + B = B + A$	$A + (B - C) = (A + B) - C$
$A + (B + C) = (A + B) + C$	$A - (B + C) = (A - B) + C$

1. **We know that the equations on the left are true for all integers and fractions.**
   Do you remember the names of any of these properties? (Additive Identity, Commutative Property of Addition, Associative Property of Addition)

2. **Do you think any of the equations on the right will work for all integers and fractions?**

3. **Let's divide the room into three groups to see if we can find examples of when these equations will not work.**
   ($A - B = B - A$ and $A - (B + C) = (A - B) + C$ are not always true; $A + (B - C) = (A + B) - C$ is always true.)

Operations

Distance/Time Formulas

Set up: Students need slates or paper.

1. **A cheetah can run at a top speed of 70 mph for short distances. Let's say a cheetah ran at top speed for 5 minutes. How far did it go?** ($5\frac{5}{6}$ mi)

2. **After its 5-minute sprint, the cheetah got tired and slowed to 18 mph for the next 20 minutes.**
   How far did the cheetah travel in total? ($11\frac{5}{6}$ mi)

3. **Work with a partner to create a formula that will determine the distance a cheetah runs if you know its running time and speed. Use *S* for speed, *D* for distance, and *T* for time.** ($S * T = D$)
   Use your formula to make up and solve another running cheetah problem.

Source: Kevin Seabrooke (Ed.), *The World Almanac for Kids 2005* (New York: World Almanac Education Group, 2004), 26.

Operations

Operations

Time, Distance, Speed: Unit Analysis/Conversion

Set up: Write the following on the board:

Top Speeds: Peregrine falcon, 200 mph
Human, 23 mph
Garden snail, 0.03 mph

Students need calculators and paper or slates.

1. **On a board are the top speeds for a falcon, a human, and a snail. If a falcon flew at top speed for one minute, how far would it get?** ($3\frac{1}{3}$ mi)

2. **How long would it take a snail at top speed to cross a road that was 30 feet wide?** ($11\frac{4}{11}$ min)

3. **A human can run at top speed for only a few seconds. In 10 seconds, how far could a person get?** ($\frac{23}{360}$ mi)

Source: Kevin Seabrooke (Ed.), *The World Almanac for Kids 2005* (New York: World Almanac Education Group, 2004), 26.

Order of Operations

Set up: On the board, write 1, 2, 3, 4, 7, 8, +, –, *, ÷, (,), and the word *exponents*. Above this write 0. Then write $3^2 - (8 + 1)$. Students need slates or paper.

1. **Use any of the numbers, operations, and grouping symbols on the board to write another name for 0. I wrote an example on the board. Raise your hand if you can use five or more numbers.**
   Write the expressions on the board and have the class check them.

2. **Does anyone have an expression that uses all six numbers?**
   (Examples: $(4 \div 2 + 3)(8 - 7 - 1)$, $1^2 * (8 + 3 - 7 - 4)$)

Operations

Operations

Order of Operations

Set up: On the board, write 1, 2, 3, 4, 5, 9, +, −, *, ÷, (,), $\sqrt{\ }$, and *exponents.* Above this write 1. Then write $\sqrt{9} + 4^2 - 3 * (5 + 1)$. Students need slates or paper.

1. **Use five or six of the numbers and operations once to write another name for 1. I've written an example on the board.**
   Raise your hand if you can use all six numbers.
   Write student solutions on the board.

2. **Can you write an expression that uses all of the operations?**
   (Example: $(1 + 2)^2 * 3 \div 9 - \sqrt{4}$)
   You may not be able to use all six numbers, but use as many as you can.

Reusable: Repeat this activity, asking students for six numbers less than 10 and a target number.

Solving Proportions

Set up: Write the proportion problems under Part 1 on the board. Students need slates or paper.

1. **Copy and solve the problems on the board.**

   $\frac{1}{4} = \frac{x}{100}$ $(x = 25)$

   $\frac{2}{3} = \frac{x}{100}$ $(x = 66\frac{2}{3})$

   $\frac{5}{12} - \frac{x}{100}$ $(x = 8\frac{1}{3})$

   Discuss solutions as a class.

2. **If 15 of the 25 students in the class ate cafeteria lunch and the rest brought their lunches, what percent of the students brought their lunches?**
   Write a proportion and solve it to get the answer. $(\frac{10}{25} = \frac{x}{100}; x = 40; 40\%)$

Operations

Finding Percents

Set up: Write the following on the board: Olympic Medals Won: Total, 929; U.S., 103; Russia, 92; China, 63. Students need slates or paper and calculators, if available.

1. **The U.S., Russia, and China won the most medals at the 2004 Summer Olympic Games.**
   About what percent of the total medals did the U.S. win? (about 11%)

2. **What percent of the medals did China win?** (about 7%) **Russia?** (about 10%)

3. **All together, about what percent of all the medals did the top three winning countries win?** (about 28%)

Comparing Rates

Set up: Draw the chart below on the board. Students need slates or paper.

Distance in Miles

	1 hour	2 hours	3 hours	4 hours	10 hours
Car A: 40 mph	(40)	(80)	(120)	(160)	(400)
Car B: 60 mph	(60)	(120)	(180)	(240)	(600)

1. **Two cars began traveling at the same time, from the same location, but in directly opposite directions. One car drove at 40 mph and the other car drove at 60 mph. Copy the chart on the board and fill it in.**

2. **Make a chart showing the distance between the two cars after 1, 2, 3, 4, and 10 hours.**

Distance Apart	(100)	(200)	(300)	(400)	(1,000)

3. **After one hour the cars are 100 miles apart, and after 10 hours they are 1,000 miles apart. Is this realistic? Why or why not?** (No, most cars can't travel at a constant speed for 10 hours.)

Operations

Data Landmarks

Set up: Write the numbers 10, 7, 23, 5, and 10 on the board.

1. **What is the median of this group of numbers?** (10)
   What is the range? (18)
   What is the mean? (11)

2. **Predict how the mean will change if 23 is removed from the data set.**
   (It will decrease.)
   Mentally calculate the new mean. (8)
   What is the new median? (8.5) **Explain why it changed.**

3. **If you think a statement is true, show me a thumbs-up. If you think a statement is not true, show me a thumbs-down.**
   Sometimes the mean is more than the maximum. (false)
   The mean may be one of the numbers in the data set. (true)
   The mean must be one of the numbers in the data set. (false)

Median and Mean

Set up: Students need calculators and slates or paper.

1. **Determine your age in months and write it down.**
   Let's take a sample of five ages.
   Ask five students to report their ages in months and record them on the board.

2. **Display the data in a stem-and-leaf plot on your paper.**
   What is the median? Use your calculator to determine the mean.

3. **Let's add five more pieces of data.**
   Have five different students provide their ages.
   Tell me how to display the data in a stem-and-leaf plot on the board.
   Now that we have new data, do you think we have a different median?
   Why? What is the median?
   Will we have a different mean? Why?
   What is the mean?

4. **Now let's add data and find the median and mean for the whole class.**

Data

Stem-and-Leaf Plot and Landmarks

Set up: Sketch the stem-and-leaf plot on the board. Students need calculators, if available.

Stems (10 s)	Leaves (1s)
2	1 1 3
3	0 5 7
4	6

1. **This stem-and-leaf plot shows the minutes it took a group of students to get to school.**
   What is the range of traveling time to school? (25)
   What is the minimum? (21) **The maximum?** (46)
   What is the median? (30)

2. **Do you think that the mean will be greater than, less than or about equal to the median? Why?** (Example: About the same because the 3 small times balance the 3 large ones.)

3. **Explain how to find the mean.** (Add up the numbers and divide by 7.) **Predict the mean and check your prediction.** (about 30; 30.43)

Data

Data with Negative Numbers

Set up: Draw the chart below on the board. Students need calculators and slates or paper.

J	F	M	A	M	J	J	A	S	O	N	D
–18	–14	–2	20	38	49	52	47	36	18	–6	–15

1. **Copy this chart. This is the average lowest temperature in degrees Fahrenheit for each month in Fairbanks, Alaska. Records have been kept for over 30 years. What are the minimum and maximum of this data set?** (–18, 52) **What is the range?** (70) **How did you determine the range?** (Subtract the minimum from the maximum.)

2. **What is the median?** (19) **What is the mean?** (about 17)

3. **Which landmarks of this data will be most useful to someone who is moving to Fairbanks and is wondering how cold it will be throughout the year? Why?** (Example: The maximum and the minimum, because it will allow him or her to be prepared for all possibilities.)

Source: University of Utah Department of Meteorology, *www.met.utah.edu/jhorel/html/wx/climate/mintemp.html*

Data

Circle Graphs

Set up: Write the following on the board: Programs (4), Decoration (6), Food (8), Invitations (2) and Chairs (4). Students need slates or paper.

1. **Students asked to put on a parent show. All students signed up to be on one committee.**
   What percent of students are working on each committee? (Programs, 16.7%; Decoration, 25%; Food, 33.3%; Invitations, 8.3%; Chairs, 16.7%)
   Sketch a rough circle graph for the data. Label each section and give the graph a title.

2. **I'm going to read some statements. If a statement is true, show me a thumbs-up. If it is false, show me a thumbs-down.**
   The program and food committees include half of the students. (true)
   One-fourth of the students are on the chair committee. (false)
   Less than one-third of the students are working on decorations. (true)
   The invitation committee is less than 10% of the class. (true)

Graphs: Actions over Time

Set up: Draw a set of coordinate axes on the board, labeled as below. Encourage students to use their slates or paper and draw graphs.

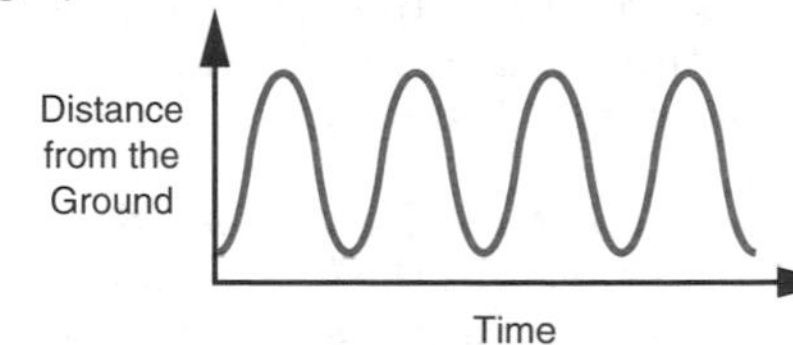

1. **Let's draw a graph of a person swinging on a swing. Our graph should show how close to and away from the ground the person goes as time passes. How should we start?** Draw exactly what students tell you and let the group correct errors until the graph looks similar to the illustration.

2. Draw a clean set of axes on the board, labeled the same way. **Describe how to draw this situation: A child climbs up a ladder and slides down a slide. Then the child does it again.**

3. **Let's continue on the same graph. Now the child climbs up some steps and crawls through a tunnel with a bump in the middle. At the end of the tunnel the child jumps to the ground.**

Reusable: Repeat this activity with different situations. Let students describe various scenarios.

Time Graph

Set up: Sketch two axes on the board. Students need slates or paper.

1. **We are going to sketch a graph that shows Lauren's speed as she ran at the park. How should we label the axes?** ("Time" and "Speed") **Draw and label the horizontal and vertical axes on your paper.**

2. **Lauren stood for a minute and then started to run. Where should the line start?** (at the origin) **Show me with your hand whether the line stays flat, slants up, or slants down.** (Stays flat and then slants up) Have students sketch a small portion of the graph on their papers.

3. **Here's what Lauren did: She ran across a flat playing field. Then she struggled up a steep hill. At the top she stopped and caught her breath. Then she flew down the hill and flopped on the grass. Continue the graph to show Lauren's speed as she exercised.** Review the story of Lauren's run as students are sketching.

4. **Why does the graph line go down when Lauren is running up the hill and up when Lauren is running down the hill?** (Example: She would slow down when she was running uphill and speed up when she was going downhill.)

Circle Graph

Set up: Students need slates or paper. Write the following on the board: African penguins, 30; rockhoppers, 13; little blues, 12.

1. **In 2005 there were three species of penguins living together in the New England Aquarium penguin exhibit. Which kind of penguin do you think a visitor would most likely see first?** (an African penguin) **Why?** (There are more of them than the other kinds.) **Which species might be the hardest to see?** (little blues) **Why?** (They have the fewest numbers.)

2. **Talk to a partner and estimate about what percent of the penguins belong to each species.** (African penguins, 50%; rockhoppers, 25%; little blues, 25%) **Sketch a circle graph showing the penguin population. Label each section. Describe how you decided to divide the circle.** (Example: I gave the African penguins a little more than half of the circle, and I gave the rockhoppers a little more than half of what was left. The little blues got the rest of the circle.)

3. **How would I change the graph if the Aquarium loans 8 African penguins to another aquarium?** (Example: The African penguin section would get smaller and the other two sections would get larger.)

Data

Probability: Counting

Set up: Draw a circle and a square on the board. Give one student a coin and another a die.

1. Have the first student flip the coin and the second student roll the die. Record the coin result in the circle (H or T) and the die result in the square. **What is the probability of getting this result when you toss a coin?** ($\frac{1}{2}$ or 50%) **What is the probability of getting this result when you roll a die?** ($\frac{1}{6}$) **If I toss this coin and roll this die again, what is the probability that I will get this result?** ($\frac{1}{12}$)
   Discuss how to determine this probability using the Multiplication Counting Principle or tree diagrams. Have students illustrate these methods on the board.

2. Write a new coin toss and die roll combination, such as H and 2, on the board. **What is the probability that I will get this result?** ($\frac{1}{12}$)
   Discuss the idea of fairness and equally likely outcomes.

3. **What is the probability that I toss a HEADS and roll an even number?** ($\frac{1}{2} * \frac{3}{6} = \frac{1}{4}$)

Vocabulary Review

Set up: Write the terms Experiment, Outcomes, and Event on the board.

1. **Give me some examples of situations where we would talk about the probability of something.** (Example: Tossing a coin)

2. **Look at the three terms on the board and listen to the following scenario:**
   We toss a coin and roll a die. We want to know the probability of getting HEADS and an even number.
   What is the experiment? (Toss one coin and roll one die)
   What are the possible outcomes? (H1, H2, . . ., H6, T1, T2, . . ., T6)
   What is the event we are looking for? (HEADS and an even number)
   How can our event occur? (By tossing a HEAD and rolling a 2 or 4 or 6)
   What is the probability of our event occurring? ($\frac{1}{4}$)

Reusable: Repeat this activity using a variety of experiments and events. You can extend the activity with a second coin or second die.

Probability

Doing an Experiment: Coin Toss–Die Roll

Set up: Students work in groups of at least three. Each group has one coin and one die. Draw the tally chart below on the board.

	1	2	3	4	5	6
H						
T						

1. **What is the probability of getting a 2 when I roll a die? A 5? A 3?** (all $\frac{1}{6}$)
What are the outcomes of tossing a coin? (H or T)
What is the probability of getting HEADS or getting TAILS? (both $\frac{1}{2}$)
So, the outcomes of both of these experiments are equally likely.

2. **If I toss a coin and roll a die, how many different results are possible?** (12) **Are they all equally likely?** (yes)
Let's do the experiment. In your group have one person toss your coin, one roll your die, and a third record your results. Do the experiment of one toss and one roll six times. When you are done, tell me your results.

3. **Did any group have six totally different combinations? Why not? Let's look at the class results. Is this what we should have expected?**

Probability

Multiplication Counting Principle

Set up: Draw a box on the board. Inside the box draw five circles labeled 1, 2, 3, 4, 5.

1. **Some lotteries use numbered balls inside a box. A machine randomly selects one ball at a time to determine the winning lottery number. Let's say we want to start a lottery using the box and balls shown on the board. Without looking, we select 2 balls, the first for the 10s and the second for the 1s. If we select the #2 ball first and then the #5 ball, the winning number is 25. What is the largest number that could be a winner?** (54) **The smallest?** (12) **List 3 numbers that cannot be winners.** (Examples: 17, 26, 49)

2. **How would we determine how many different 2-digit number winners are possible?**
   Record student ideas on the board. Include the methods "carefully make a list," "use a tree diagram," and "use the Multiplication Counting Principle."
   Which is the fastest method to find the number of 2-digit possibilities? How many possible winning numbers are there? (20)

3. **How many possible winning numbers would there be if we had a 3-digit lottery?** (60) **A 4-digit lottery?** (120) **A 5-digit lottery?** (120)

Probability

Probability

Tree Diagrams

Set up: Students need slates or paper. Write the following on the board: Bread: White or Wheat; Meat: Ham, Turkey or Salami; Cheese: American or Swiss

1. **Draw a tree diagram to show all the sandwiches that can be made with these choices. How many sandwiches are possible?** (12 sandwiches) **How would you adjust your diagram if there's a choice of mayonnaise or mustard?** (Add a row at the bottom.) **We can't find the probability of a particular sandwich being *ordered*—that's not a random process—but we can find the probability of a sandwich being selected at random. From our tree diagram, can we find the probability that the randomly selected sandwich is a ham and Swiss on wheat with mustard?** (yes, $\frac{1}{24}$)

2. **Let's do another tree diagram. We pick two balls from a box containing four balls marked with the numbers 1 through 4. The first ball picked is the 10s digit and the second is the 1s digit. How many two-digit numbers are possible?** (12) **From our tree diagram, can we find the probability of getting the number 31?** (yes, $\frac{1}{12}$)

Probability: Counting

Set up: Draw the illustration below on the board.

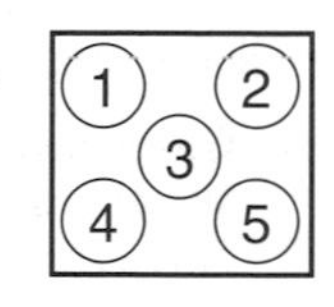

1. **I have five balls numbered 1 to 5 in a box. Without looking I pick one. What is the probability that I will pick the #2 ball? The #5 ball? The #1 ball?** (all $\frac{1}{5}$) **So, we can say that all of our results are equally likely.**

2. **Let's say we picked the #1 ball, but we do not put it back into the box.** You can erase or cross out the #1 ball in the box and write a 1 next to the box. **What is the probability that I will now pick the #2 ball? Or the #4 ball?** ($\frac{1}{4}$)

3. **Let's start again with five balls in the box. Without looking we pick two balls, one at a time. As we pick each ball we record its number to form a two-digit number. The first ball is the tens digit and the second is the ones digit. How many two-digit numbers are possible?** (20)
   What is the probability that I pick 12? ($\frac{1}{20}$)

4. **Here's a challenge: Suppose I pick two balls, but I don't care about the order. So, 1 and 2 is the same as 2 and 1.**
   How many different two-number combinations can I pick? (10)

Probability

Area

Set up: Students need slates or paper and a ruler or measuring tape. Fraction calculators are optional.

1. **Calculate the area of a piece of notebook paper.** (93.5 in^2)
   What unit should we use for the area of our paper? (square inches)
   Have students explain how they did their work.

2. **Using this information, estimate the area of your desk and then verify by measuring.**
   Discuss solutions and units.

3. **Using what you think is the best estimate of desk area, calculate the total surface area for all of the desks in the room.** (Desk area * number of desks)

Measurement

Area of Circles

Set up: Students need slates or paper. Calculators and grid paper are optional.

1. **Rockport House of Pizza sells a round 10-inch cheese pizza for $5.25 and a round 16-inch pizza for $8.25. Which pizza is a better buy?**
   How do you think we should approach this problem? (Calculate the cost per square inch of each pizza.)
   Work with your partner to determine the better buy. (16-inch pizza)

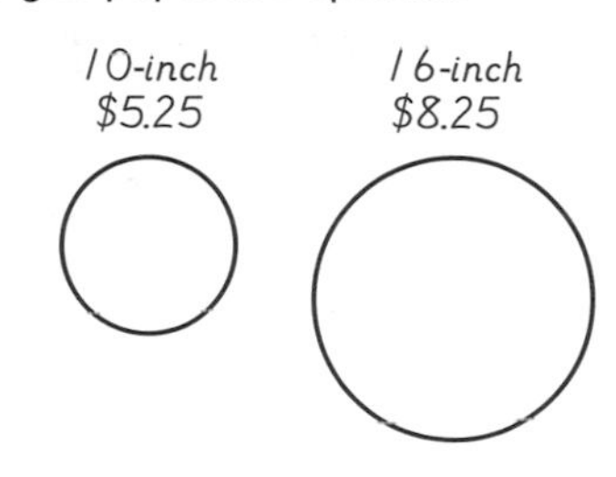

2. **The owners are thinking of offering rectangular pizzas for larger groups.**
   Given what you already know about their pricing, what sizes and prices should they consider making?
   Have selected students present their ideas.

Measurement

Volume

Set up: Draw the shapes and the chart below on the board. If possible, have a sample of each shape available. Students need slates or paper.

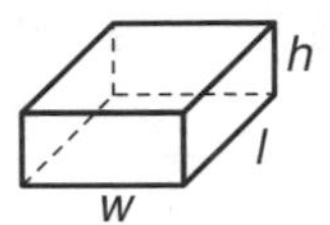

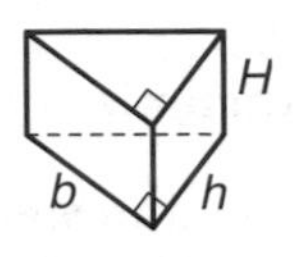

Shape	Area of Base	Volume
Rectangular Prism	($A=lw$)	($V=lwh$)
Triangular Prism	($A=\frac{1}{2}bh$)	($V=\frac{1}{2}bhH$)
Cylinder	($A=\pi r^2$)	($V=\pi r^2 h$)

1. **Draw this chart on your paper and write in the formulas for the area of the base and the volume of each shape.**
   Go over the formulas when most students seem to have them written.

Measurement

continued

2. Draw the second chart and label the dimensions on the sketches on the board as follows: $l = 2$, $w = 4$, $h = 5$, $r = 3$, $b = 2$, $H = 5$.
Find Volume 1 for each shape.

Shape	Volume 1	Volume 2
Rectangular Prism	(40 in^3)	(320 in^3)
Triangular Prism	(25 in^3)	(200 in^3)
Cylinder	(141.3 in^3)	(1,130.4 in^3)

3. **Let's now double the dimensions on each shape and calculate Volume 2. How do the volumes differ when their dimensions are doubled?** (They are multiplied by 8 or 2^3.)

Measurement Comparison

Set up: Draw the chart below on the board. Students need slates or paper.

Attribute	Smallest Unit			Largest Unit		
	U.S. Customary	><	Metric	U.S. Customary	><	Metric
Length	(inch)	(>)	(millimeter)	(mile)	(>)	(kilometer)
Weight	(ounce)	(>)	(milligram)	(ton)	(<)	(metric ton)
Capacity	(teaspoon)	(>)	(milliliter)	(gallon)	(>)	(liter)
Volume	(cubic inch)	(>)	(cubic centimeter)	(cubic yard)	(<)	(cubic meter)

1. **Fill in the chart with the smallest commonly used units for each attribute in each system. For each attribute, indicate which type of unit of measure is bigger by using the less-than or greater-than sign.**

2. **Now fill in the chart with the largest units used to measure each attribute. For each attribute, use the less-than or greater-than sign to indicate which of the two types of unit measures is bigger.**

Time Conversions: Percent Application

Set up: Write the following statement on the board with times for your school day: School begins at _________ and ends at _____________. Students need slates or paper and calculators.

1. **How many minutes are you in school each day?**

2. **What percent of the minutes in a day are you in school?**

3. **How many minutes are there in a week?** (10,080 min)

4. **What percent of the week are you in school?**

5. **What percent of the year are you in school?**

Measurement

Scale

Set up: Write "1 inch stands for 1 foot" on the board. Students need slates or paper.

1. **Write down your approximate height in feet and inches and then in inches. How tall would you be in a drawing if you use this scale?**

2. **A canoe is 15 feet and 8 inches long.**
   Using the 1 inch to 1 foot scale, how long would a model canoe be?
   ($15\frac{2}{3}$ in. or 1 ft $3\frac{2}{3}$ in.)

3. **Model paddles are $6\frac{1}{2}$ inches long.**
   How long are the real paddles? (6 ft 6 in. or $6\frac{1}{2}$ ft)

Source: Old Town Canoe Co., *www.oldtowncanoe.com/canoes_charles.php*

Measurement

Scale

Set up: Draw the illustration below and write "1 unit stands for 1 yard" on the board. Students need slates or paper.

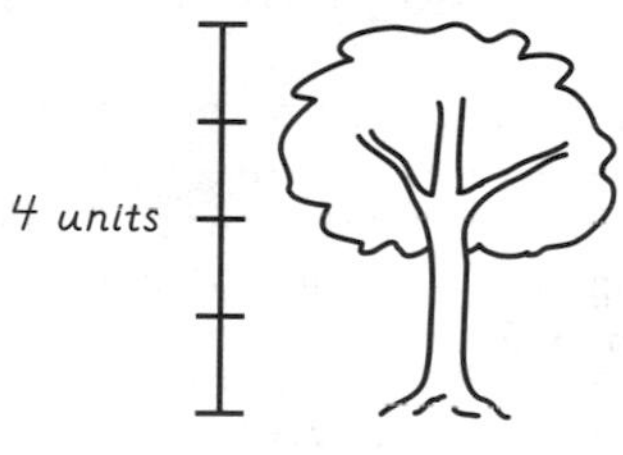

1. **Here is a scale drawing of a tree. Each unit represents one yard. What is the height of the tree in yards?** (4 yd) **In feet?** (12 ft)

2. **Use the same scale to make a drawing of a tree that is 10 feet tall.**

3. **Draw yourself to scale next to the tree.**

Vocabulary and Properties

Set up: The board is needed.

1. **Help me draw some shapes on the board. Who can tell me how to draw an isosceles triangle?**
   Draw exactly what the students say. For example, a student might suggest, "Draw three lines, two equal and one different." You should then draw three lines, but don't connect them into a triangle because that was not stipulated. If students have trouble guiding your drawing, remind them that they can label vertices and other parts of their figures.

2. **Tell me how to draw two concentric circles.**
   Students should be a little more articulate at this point.

3. **Tell me how to draw a trapezoid.**

4. **How should I draw a pentagon with a diagonal?**

Reusable: Repeat this activity, having students work in pairs. One student describes how to draw a particular shape to his or her partner. Without letting the first student see, the second student uses his slate or paper to draw what is described.

Geometric Relations

Set up: None.

1. **What makes two shapes similar?**
   Students should note that similar figures have the same shape, but not necessarily the same size.

2. **If two figures are similar, one is an exact copy or a proportional enlargement of the other.**
   Are all rectangles similar? Why? (No, not always a proportional enlargement)

3. **Are all circles similar?** (yes)
   Are there other figures that are all similar? (Example: All squares are similar.)

Geometry

Paper-Folding Angles: Part I

Set up: Students need a sheet of $8\frac{1}{2}$ by 11 inch paper. Students work alone or with a partner.

1. **First, we're going to make a square from the $8\frac{1}{2}$ by 11 inch paper you have. Line up the bottom edge of the paper with the lower part of the left edge like this. Help each other. Make a sharp crease. Now fold the top of the paper over like this and make another sharp crease. Cut or tear off this top piece. Now you have a square. What is the fold in the square called?** (the diagonal) **The angle it bisects is how many degrees?** (90°) **Each of the smaller angles has what measure?** (45°)

cut top off

fold

2. **Now line up the bottom edge of the paper with the diagonal fold like this and make a sharp crease. How many angles do you see in the bottom left corner made by the creases?** (3) **What is the measure of the largest angle?** (45°)
Notice the two bottom angles are the same. What is the measure of each? ($22\frac{1}{2}$°)

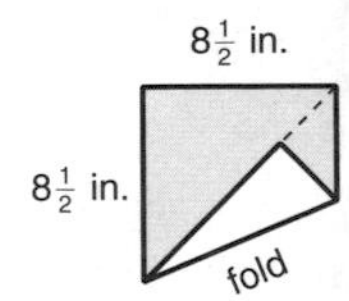

Geometry

3. **Line up the bottom of your paper with the crease that bisects the 45° angle and make a sharp crease. Repeat this folding and creasing process once more. Label the measures of the new angles created by the folds. Use either fractions or decimals.** (45°, 22.5° or $22\frac{1}{2}$°, 11.25° or $11\frac{1}{4}$°, 5.625° or $5\frac{5}{8}$°)

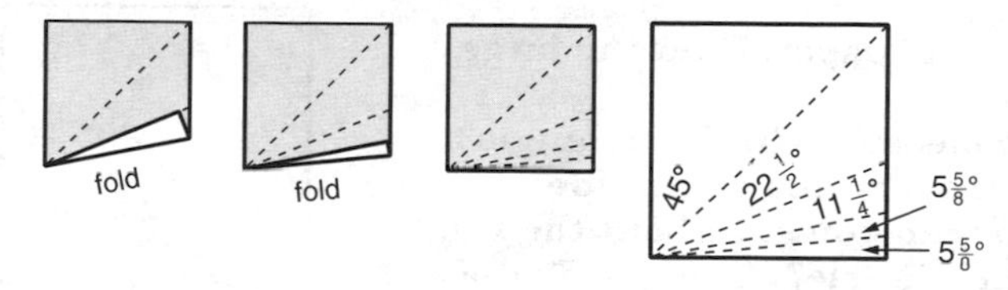

NOTE: Students should save their folded-angle papers to use with Part II.

Paper-Folding Angles: Part II

Set up: Students should do Part I before doing this activity. They should use the folded-angle papers they created during Part I or quickly make new ones. Students work alone or with a partner.

1. **Look at your squares of paper. Find the largest right triangle.**
   What are the angle measures in this triangle? (90°, 45°, 45°) **Label them on the clean side of your paper. What can you say about the side lengths of this right triangle?** (Example: Two of the sides have the same length.)

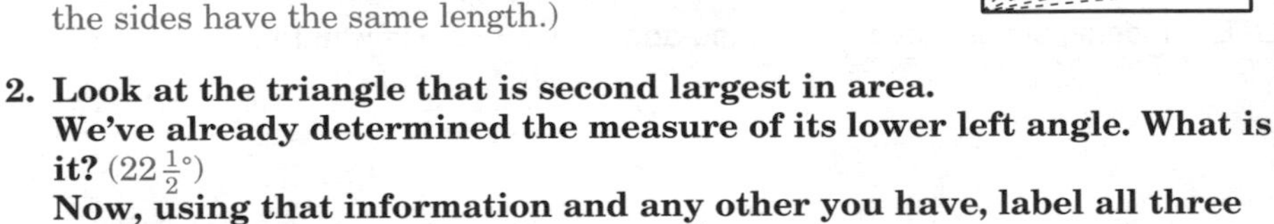

2. **Look at the triangle that is second largest in area.**
   We've already determined the measure of its lower left angle. What is it? ($22\frac{1}{2}°$)
   Now, using that information and any other you have, label all three angles in this triangle. ($22\frac{1}{2}°$, 45°, $112\frac{1}{2}°$)

3. **Do the same for the remaining triangles on your paper.**

Angles

Set up: The board is needed. Students need protractors and slates or paper.

1. **On your slate or paper, list the names for types of angles by size from smallest to largest.** (acute, right, obtuse, straight, reflex)
   List the terms offered by the students on the board.

2. **Look around our classroom and find an example of each type of angle. Estimate the size of each in degrees.**
   Ask selected students to give their ideas and to verify their guesses by measuring the angles with a protractor.

3. **Which type of angle is the most common? Why do you think this is so?**

Angles and Transversals

Set up: Draw a pair of parallel lines on the board. Students need slates or paper and a ruler.

1. **Use a ruler to draw a pair of parallel lines. Measure with the ruler to be sure the lines are parallel.**
   Draw another pair of parallel lines in a diagonal direction across the first pair. Measure with the ruler to be sure these lines are parallel.

2. **Label the points of intersection from the top left corner with the letters *A* through *D*. Estimate the size of the interior angle at *A*. Now, using what you know about angles, determine a good estimate for the size of the other angles.**

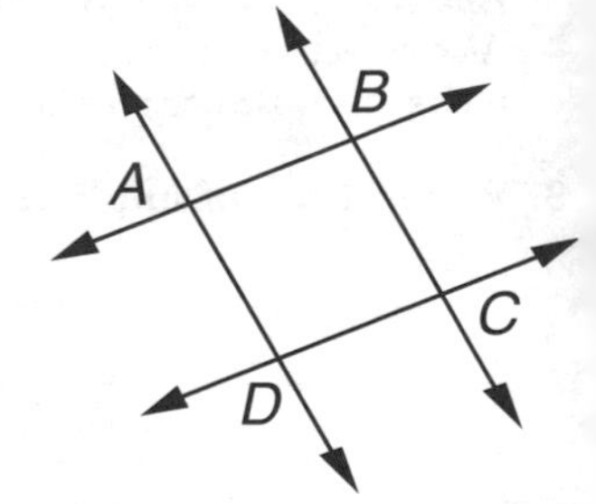

3. **How did you determine the estimate for each of the interior angles?**

Geometry

Transformations

Set up: Students need paper. Graph paper, rulers, or transparent grid paper are optional. On the board, write the ordered pairs (2,0), (2,4), (4,2), (6,4), and (6,0).

1. **Fold a sheet of paper into fourths so the folds form the axes for a graph. Darken the axes with your pencil.**
   Plot the points shown on the board and connect them in order. (It looks like a capital M.)

2. **Slide the drawing four units to the right. Write the new coordinates on the graph.** ((6,0), (6,4), (8,2), (10,4), (10,0))

3. **Flip both drawings over the horizontal axis.**
   Write the coordinates for the new figures on the graph. (First figure flipped: (2, 0), (2, –4), (4, –2), (6, –4), (6, 0); second figure flipped: (6, 0), (6, –4), (8, –2), (10, –4), (10, 0))

Intersecting Lines and Angles

Set up: Students need paper. Draw the illustration below on the board.

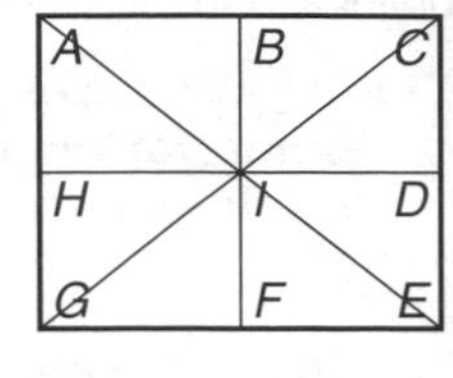

1. **Fold a piece of $8\frac{1}{2}$ by 11 inch paper in half the long way and in half again the short way. Now fold it to form the two diagonals.**
   Open the paper and beginning in the top left corner, label corners and crease points with the letters *A* through *H*, as shown on the board. Label the center point *I*.

2. **Identify some parallel and perpendicular line segments.** (Example: $\overline{AC}$ is perpendicular to $\overline{BI}$)

3. **Identify some complimentary and supplementary angles on the figure.** (Examples: $\angle HIA$ and $\angle AIB$ are complementary, $\angle DIF$ and $\angle HIB$ are supplementary)

Geometry

Parts of 3-Dimensional Shapes

Set up: Draw the table below on the board. Students need slates or paper.

Prisms

Shape of Base	Sketch	# of Faces	# of Vertices	# of Edges
Triangle		5	6	9

1. **We're going to look at characteristics of different prisms. Copy this table. Let's start with a triangular prism. We write "triangle" in the column for shape of bases. Now, sketch a picture of a triangular prism. How many faces, vertices and edges does a triangular prism have?** You might show students one easy way to draw a quick sketch by placing the two bases near each other and connecting the related vertices.

2. **Add more rows and fill in the table for different types of prisms. Compare your entries with a partner.**

3. **Describe any patterns you notice.**

Symmetry: Line of Reflection Outside the Figure

Set up: Draw $\triangle ABC$, $\overleftrightarrow{MN}$, and $\overleftrightarrow{OP}$ on the board. Students need slates or paper.

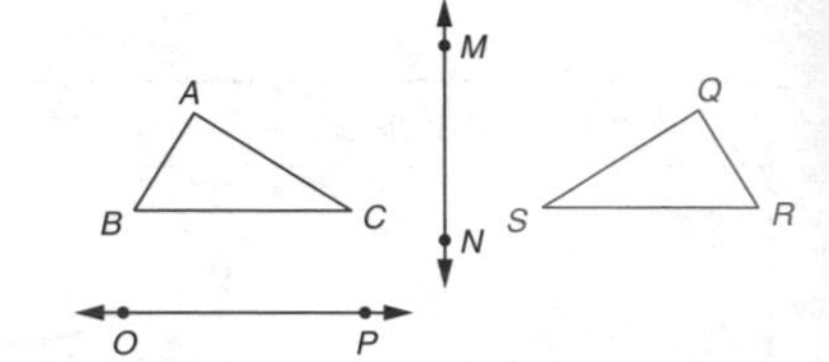

1. **Draw $\triangle ABC$ reflected in $\overleftrightarrow{MN}$.**
   Label the reflection $\triangle QRS$.
   Finish this sentence: A figure and its reflection are ______. (congruent)

2. **What point is the same distance from $\overleftrightarrow{MN}$ as A?** (Q)

3. **Mark any point of the triangle ABC with an x.**
   Now reflect $\triangle ABC$ in $\overleftrightarrow{OP}$.
   Use another x to show where the point you marked is located on the reflected image.

Geometry

Combining Like Terms

Set up: Write the problems below on the board. Students need slates or paper.

1. **Simplify each expression by combining like terms. Check each problem by substituting 4 for the variable.**
   $12m + 3m - 2m =$ ($13m$, 52)
   $(8x - x) - 4x =$ ($3x$, 12)
   $y + 3y + 7y =$ ($11y$, 44)
   $x - x + 5x =$ ($5x$, 20)
   $10x + 3x + 2 =$ ($13x + 2$, 54)
   $9y - y + 4 =$ ($8y + 4$, 36)

   Discuss student solutions.

2. **Make up three problems like these and trade with a partner. Check each other's answers.**

Using Formulas

Set up: Students need paper. Write $A = bh$ on the board.

1. **Here is the formula for the area of a rectangle. What does each letter represent?** (b: base; h: height)

2. **Write a formula you could use to determine the amount of money you would make if your aunt pays you $3.50 per hour for helping her. Use the variable *m* for money and *h* for the number of hours you work.** ($m = 3.50h$)

   Have students share their formulas. Then do one example on the board.

3. **What formula could you use to determine the number of hours you spend in school in a school year? Use *h* for the number of hours in a school year and *d* for the number of days in a school year.** (Example: $h = 8d$, if there are 8 hours in a school day)

   What formula would you use to change those hours into school days? (Example: $d = h / 8$)

Reusable: Repeat this activity, writing formulas relating to other daily activities.

Grouping Like Terms

Set up: Write the problems for #1 and #2 on the board. Students need slates or paper.

1. **Fill in the blank in each equation below. Here's an example:**
   $5y - y =$ ___. So, $4y$ goes in the blank.
   Solve the rest by grouping the variables as indicated in each problem.

$4a + a =$ ___	$(5a)$	$3b - b =$ ___	$(2b)$
___ $= 7m - 2m$	$(5m)$	$8y + 2y + y =$ ___	$(11y)$
$x + x =$ ___	$(2x)$	$11s + s =$ ___	$(12s)$

   Let's review our solutions.

2. **Here are some more problems for you to practice.**

$10s - 8s =$ ___	$(2s)$	$n + 5n - n =$ ___	$(5n)$
$7a + 2a - a =$ ___	$(8a)$	$5x + 3 + x =$ ___	$(6x + 3)$
$r - r + 7 =$ ___	(7)	___ $= 12m \div 3m$	(4)

   Check your answers with a partner.

Patterns: Positive and Negative Numbers

Set up: Draw the *in/out* table below on the board. Students need slates or paper.

1. **Work with a partner. Copy the table and complete it. Challenge each other with the blank rows.**

2. **Use words to write a rule for the *in/out* table.** (Example: The input divided by two, then divided by ten)
   Write the rule using the variable x.
   (Example: x / 20)

in	*out*
6	0.3
10	0.5
3	0.15
–1	(–0.05)
0	(0)

Algebra

Problems with Variables

Set up: Students need slates or paper.

1. **How many feet are on 7 dogs?** (28 feet) **8 dogs?** (32 feet) **9 dogs?** (36 feet)
   What is the general pattern? (Multiply the number of dogs by 4)
   How many feet are on 52 dogs? (208 feet)
   Write an expression for finding the number of dog feet if you know the number of dogs. ($4D$)
   How many dog feet are on D dogs? ($4D$)

2. **If there are 32 dog feet in a room, how many dogs are there?** (8 dogs)
   If you know the number of dog feet, how can you determine the number of dogs? (Divide the number of dog feet by 4)
   Write an expression for finding the number of dogs if there are F feet.
   ($\frac{F}{4}$)

3. **Which numbers of dog feet would be highly unlikely to occur in any group of dogs? Why?** (Any number that is not a multiple of 4, since most dogs have 4 feet)

Problems with Variables

Set up: The board is needed. Students need slates or paper.

1. **Even though the word parts in *centipede* mean "100 feet," the common house centipede has 15 pairs of legs. Write an expression for finding the number of legs on a group of house centipedes. Use C for the number of centipedes.** ($30C$)
   If there are C centipedes, how many legs are there? ($30C$)

2. **Use your rule to determine which numbers of centipede legs will never occur. Why won't the numbers occur?**

3. **The museum has octopuses and centipedes on display.**
   Write an expression for finding out the total number of octopus legs and centipede legs at the museum using P for the number of octopuses and C for the number of centipedes. ($8P + 30C$)

Source: Enchanted Learning,
www.enchantedlearning.com/subjects/invertebrates/arthropod/Centipede.shtml

Algebra

Writing and Evaluating Algebraic Expressions

Set up: Students need slates or paper. Write $\frac{x}{2}$ on the board.

1. **What is the x called?** (a variable)
   Work with a partner to write two different stories that fit the expression $\frac{x}{2}$. For example, "David is half as old as his brother."

2. **Write an algebraic expression for each of these statements:**
   A teacher is x years older than her 12-year-old students. ($x + 12$)
   Aisa's school year was k weeks long and she checked out 3 library books every week. One week she was sick and missed library.
   ($3k - 3$ or $3(k - 1)$)
   Brenda shares her package of licorice evenly among S friends. ($\frac{1}{S}$)

3. **Write an expression for this statement: Each student in our class was given m pencils and 5 pencils were given to the teacher.**
   Calculate the total number of pencils if $m = 3$.

Algebra

Multiplying by Unit Fractions

Set up: Write 2 * ___ = 1 on the board. Students need slates or paper.

1. **Read this sentence. What number will make this statement true?** ($\frac{1}{2}$)

2. Write 2 ÷ ___ = 1 on the board.
   Now, if we divide instead of multiply, what number goes in the blank? (2)

3. Now write these problems on the board.
   Work with a partner and try these problems.

14 * ___ = 7	($\frac{1}{2}$)	5 * $\frac{1}{2}$ = ___	(2.5)
14 ÷ ___ = 7	(2)	5 ÷ 2 = ___	(2.5)
9 * ___ = 3	($\frac{1}{3}$)	5 * ___ = 1	($\frac{1}{5}$)
9 ÷ ___ = 3	(3)	5 ÷ ___ = 1	(5)

4. Finally, write these problems on the board.
   Finish these sentences.

$A * \frac{1}{2}$ = ___	($\frac{A}{2}$)	$A * \frac{1}{3}$ = ___	($\frac{A}{3}$)	$A * \frac{1}{N}$ = ___	($\frac{A}{N}$)
$A \div 2$ = ___	($\frac{A}{2}$)	$A \div 3$ = ___	($\frac{A}{3}$)	$A \div N$ = ___	($\frac{A}{N}$)

Integer Sums and Differences

Set up: Sketch a two-column table on the board. Label the columns *x* and *y*. Students need slates or paper.

1. **Tell me a pair of integers whose sum is −7.** (Example: 7 and −14)
   Name other values of x and y that have a sum of −7.
   Fill in the chart as pairs are suggested.
   As x decreases, what happens to y? (It increases.) **How many values of x and y have a sum of −7? Explain.** (An unlimited number)

2. Draw a new two-column chart on another section of the board.
   Tell me a pair of whole numbers x and y such that $x - y$ equals −7.
   (Example: 2 and 9)
   Work with a partner to find several pairs of integers whose difference is −7. Look at the pairs you found. As x increases, what happens to y? (It increases.) **How many pairs of whole numbers have a difference of −7? Explain.** (An unlimited number)

Reusable: Repeat this activity with numbers less than 12 for the sum and difference.

Sequence

Set up: Write..., ___, 5.9, 5.7, 5.5, ___, 5.1, ___,...on the board. Students need slates or paper.

1. **April thought of a rule and then wrote this sequence of numbers. What do you think her rule was? Explain.** (Example: Subtract 0.2) **Can you find another rule that would also work?** (Example: Add –0.2)

2. **What are the missing numbers in this sequence? Explain.** (6.1, 5.3, 4.9)

3. **Will 4.8 be a number in the sequence? Explain.** (no)

Reusable: Repeat this activity with any sequence and missing numbers in the sequence.

Writing Equations from Stories

Set up: Students need slates or paper. They can work alone or with a partner.

1. **One week Maria made some money babysitting. The next week she made $15. All together she made $33 dollars.**
   How much did she make the first week? ($18)
   Write an equation that describes this story. Use m as a variable for money. (Example: $m + 15 = 33$, where m stands for money made first week)

2. **The third week Maria babysat again. During the fourth week she made twice as much as she did in the third week. For weeks three and four she made a total of $60.**
   Write an equation for this story and solve it. ($x + 2x = 60$, where x stands for amount of money made in third week; $x =$ $20)

3. **After her fifth week of babysitting, Maria went shopping and spent $\frac{2}{3}$ of the money she made that week.**
   If she ended up with $9, how much had she made that week? ($27)
   Write an equation for the story and solve it. (Example: $\frac{1}{3}y = 9$; $y = 27$, where y is the amount of money made in the week)

Sequence: Square Numbers

Set up: Write 0, 3, 8, 15, . . . on the board. Students need slates or paper.

1. **Jessica thought of a rule and then wrote this sequence of numbers. What do you think her rule was? Explain.** (Example: Add the next odd number)
   Can you find another rule that would also work? (Example: Subtract 1 from the next square number)

2. **What are the next three numbers in this sequence? Explain.** (24, 35, 48)

3. **Work with a partner to find the tenth number in the sequence.** (99)

Function Machine: $n^2 - 1 = m$

Set up: Write $n^2 - 1 = m$ on the board. Draw the *in/out* table below on the board. Students need slates or paper.

1. **The natural numbers 1, 2, 3, and so on are put into a function machine in order. Here is the rule for the machine. What is the first number that will go into the machine?** (1) **What is the first number that will come out?** (0) **Write the first five numbers that will come out of the machine in order.** (0, 3, 8, 15, 24)

in: n	*out: m*

2. **What is the tenth number that will come out?** (99)

3. **Will 144 come out? Explain.** (No, 144 is a square number.)

Algebra

Functions

Set up: Draw the chart below on the board. Students need slates or paper.

x	y
0	–1
(2)	3
4	(7)
(6)	11
8	15
(1)	1

1. **Copy the chart on the board and fill in the missing values.**

2. **Write an equation using x and y that shows the rule you used to complete the chart.** (Example: $y = 2x - 1$)

3. **Add four more rows to the chart and fill them in.**

Number Line

Set up: Draw a blank number line on the board. List the numbers 5, $-\frac{1}{2}$, 0, -2, 3, and -4 on the board. Students need slates or paper.

1. **Draw a blank number line like this one on your paper. Write these numbers in the appropriate places on your number line.**
   Fill in the number line on the board using student suggestions.

2. **Let's use our number line to do some addition and subtraction problems.**
   Use the number line to show the following addition and subtraction problems. You may want to use two colors of chalk to indicate addition (moving along the line in a positive direction) and subtraction (moving in a negative direction).

   $-1 + 5$ $\quad$ $3 - (-2)$ $\quad$ $0 + 2$ $\quad$ $\frac{1}{2} + (-\frac{1}{2})$ $\quad$ $4 - 6$

Formula

Set up: Write $P = 2 * (L + W)$ on the board. Students need slates or paper.

1. **This formula is used to find the perimeter of a rectangle. *P*, *L* and *W* are variables.**
   What do you think the *P* represents? (perimeter)
   What does *L* stand for? (length)
   What does *W* stand for? (width)
   What is the perimeter of a rectangle 8 feet long and 3 feet wide? (22)
   7 feet long and 4 feet wide? (22)
   2 feet long and 9 feet wide? (22)
   Why do these rectangles all have the same perimeter?

2. **Work with a partner to find the dimensions of more rectangles that have a perimeter of 22 feet.**

ist of Activities by Page

asy Activities 1

Numeration **1**

Fractions 1
Powers of Ten 2
Roman Numerals: I–XXX 4
Roman Numerals: I–XXX (Review) 6
Using Exponents 7
Fractions: $\frac{1}{2}$ 8
Divisibility Rules for 2 10
Fractions: <1, $=1$, >1 11
Whole Number Place Value 12
Whole Number Factors 13
Decimal Place Value 14
Whole Number Rounding 15
Fractional Parts 16
Equivalent Fractions 17
Whole Number Place Value 18

Operations **19**

Multiplication Estimation 19
Division Word Problems 20
Whole Number Addition 21
Fraction Multiplication 22
Fraction Multiplication 23
Whole Number Subtraction 24
Whole Number Division 25
Fraction Addition and Subtraction .. 26
Mixed Operations 27
Division Notation 28
Rates 29
Multiple Operations 30
Multiple Operations 31
Multiple Operations: $n * 0 = 0$ 32
Addition/Subtraction 33

Data **34**

The Median 34
Time Graph 35
Tally Chart 36
Mean, Median 37

Mean....................38
Reading a Bar Graph....................39
Data Landmarks....................40
Mean, Median....................41
Probability....................42
Simple Probability....................42
Probability: Range....................43
Vocabulary: Experiments, Outcomes 44
Probability Representation....................45
Doing an Experiment....................46
Probability of Events....................47
Measurement....................48
Metric Units....................48
Metric Length....................49
Perimeter....................50
Relations: Area of a Triangle....................51
Volume of a Prism....................52
Angle Measure....................53
Scale....................54
Scale Drawings....................55
Geometry....................5
Geometric Properties....................5
Geometric Vocabulary....................5
Solid Shapes....................5
Angle Classification....................5
Scale Drawing a Rectangle....................6
Properties of 2-Dimensional Shapes....................6
Symmetry....................6
Naming Line Segments....................6
Identifying Characteristics of 2-Dimensional Shapes....................6
Symmetry: Line of Reflection Outside the Figure....................6
Algebra....................6
Creating Number Sentences....................6
Multiplication Patterns....................6
Patterns....................6
Patterns....................6
Writing Algebraic Expressions....................7
Multiplying by $\frac{1}{2}$....................7

Whole Number Sums and
Differences ... 72
Formula ... 73
Sequences ... 74
Sequences ... 75
Sequences ... 76
Grouping Symbols ... 77
Using Parentheses ... 78

Moderate Activities ... 79

Numeration ... 79

Fractions ... 79
Powers of Ten ... 80
Roman Numerals: I–LII ... 82
Roman Numerals: I–L (Review) ... 84
Scientific Notation ... 85
Divisibility Rules for 3 ... 86
Fractions: $\frac{1}{2}$ and $\frac{1}{4}$ ... 87
Fractions: <1, $=1$, >1 ... 88
Decimal Place Value ... 89
Whole Number Factors ... 90
Whole and Decimal Number
Rounding ... 91
Whole Number Rounding ... 92
Fraction/Decimal/Percent
Conversion ... 93
Decimal Place Value ... 94

Operations ... 95

Whole Number Estimation,
Multiplication, Division,
Measurement ... 95
Whole Number Division ... 96
Whole Number Division ... 97
Fraction Addition ... 98
Fraction Addition ... 99
Integer Addition and Subtraction ... 100
Fraction Subtraction ... 101
Decimal Addition ... 102
Percent ... 103
Decimal Addition ... 104
Rates ... 105
Order of Operations ... 106

Multiple Operations 107
Square Root Sign 108
Rates 109
Ratios 110
Proportions 111
Fractions 112
Fraction Addition 113
Data 114
Graphs: Actions over Time 114
Data Landmarks 116
Reading Bar Graphs 117
Circle Graphs 118
Median and Mean 119
Stem-and-Leaf Plot 120
Data with Negative Numbers 121
Time Graph 122
Circle Graph 123
Reading a Time Graph 124
Probability 125
Probability: Outcomes 125
Probability: Likelihood 126
Vocabulary: Experiments,
Outcomes, Events 127
Doing an Experiment 128
Tree Diagram 129
Multiplication Counting Principle .. 130
Measurement 131
Conversion of Units 131
Capacity 132
Conversions: Yards, Feet, Inches ... 133
Units of Weight: Conversions 134
Circumference and Area 135
Time Conversions 136
Scale and Conversion 137
Scale and Perimeter 138
Geometry 139
Congruence 139
Angles in Triangles and
Quadrangles 140
Regular Polygons 141
Paper-Folding Angles 142
Volume to Surface Area 144

Classifying Quadrilaterals 145
Classifying Triangles 146
Properties of 3-Dimensional Shapes 147
Transformations 148
Symmetry: Line of Reflection Outside the Figure 149
Algebra **150**
Writing Algebraic Expressions 150
Equivalent Expressions 151
Problems with Variables 152
Patterns 153
Number Patterns 154
Whole Number Sums and Differences 155
Formula 156
Sequence: Square Numbers 157
Sequence 158
Function Machine: $\frac{n}{2} + 1$ 159
Functions 160
Number Line 161
Patterns 162
Grouping Symbols 163
Associative Property 164
Difficult Activities **165**
Numeration **165**
Fraction Equivalents, Adding Fractions 165
Powers of Ten: Numbers Less Than or Equal to 1 166
Roman Numerals: I–CXV 168
Roman Numerals: I–CXV (Review). 170
Scientific Notation: Very Small Numbers 171
Divisibility Rules for 3 and 9 172
Ordering Fractions 173
Fractions: <1, =1, >1 174
Negative and Positive Integers 175
Integer and Fraction Properties 176

Whole Number Theory 177
Decimal Rounding and Comparing 178
Percent: Mental Calculation 179
Ordering Fractions 180
Fractions: Converting to Decimals ... 181
Operations 182
Estimation, Multiplication, Percent 182
Division Word Problems, Decimals . 183
Fraction Addition and Subtraction Word Problems 184
Fraction Multiplication and Division 185
Decimal Multiplication 186
Averaging Percents 187
Fraction Division 188
Properties of Integers 189
Properties of Addition (and Non-Properties) 190
Distance/Time Formulas 191
Time, Distance, Speed: Unit Analysis/Conversion 19
Order of Operations 19
Order of Operations 19
Solving Proportions 19
Finding Percents 19
Comparing Rates 19
Data 19
Data Landmarks 19
Median and Mean 19
Stem-and-Leaf Plot and Landmarks 20
Data with Negative Numbers 20
Circle Graphs 20
Graphs: Actions over Time 20
Time Graph 20
Circle Graph 20
Probability 20
Probability: Counting 20
Vocabulary Review 20